WORLD HEALTH ORGANIZATION

CORRIGENDUM

ENVIRONMENTAL HEALTH CRITERIA

NO. 106

BERYLLIUM

Page 169:

Insert the attached page

(Gummed back permits direct attachment to incorrect page)

This report contains the collective views of an international group of experts and does not necessarily represent the decisions or the stated policy of the United Nations Environment Programme, the International Labour Organisation, or the World Health Organization.

Environmental Health Criteria 106

BERYLLIUM

Published under the joint sponsorship of
the United Nations Environment Programme,
the International Labour Organisation,
and the World Health Organization

World Health Organization
Geneva, 1990

The International Programme on Chemical Safety (IPCS) is a joint venture of the United Nations Environment Programme, the International Labour Organisation, and the World Health Organization. The main objective of the IPCS is to carry out and disseminate evaluations of the effects of chemicals on human health and the quality of the environment. Supporting activities include the development of epidemiological, experimental laboratory, and risk-assessment methods that could produce internationally comparable results, and the development of manpower in the field of toxicology. Other activities carried out by the IPCS include the development of know-how for coping with chemical accidents, coordination of laboratory testing and epidemiological studies, and promotion of research on the mechanisms of the biological action of chemicals.

WHO Library Cataloguing in Publication Data

Beryllium

(Environmental health criteria ; 106)

1.Beryllium I.Series

ISBN 92 4 157106 3   (NLM Classification: QV 275)
ISSN 0250-863X

© World Health Organization 1990

Publications of the World Health Organization enjoy copyright protection in accordance with the provisions of Protocol 2 of the Universal Copyright Convention. For rights of reproduction or translation of WHO publications, in part or *in toto*, application should be made to the Office of Publications, World Health Organization, Geneva, Switzerland. The World Health Organization welcomes such applications.

The designations employed and the presentation of the material in this publication do not imply the expression of any opinion whatsoever on the part of the Secretariat of the World Health Organization concerning the legal status of any country, territory, city or area or of its authorities or concerning the delimitation of its frontiers or boundaries.

The mention of specific companies or of certain manufacturers' products does not imply that they are endorsed or recommended by the World Health Organization in preference to others of a similar nature that are not mentioned. Errors and omissions excepted, the names of proprietary products are distinguished by initial capital letters.

Computer typesetting by HEADS, Oxford OX7 2NY, England

Printed in Finland
90/8657 — Vammala — 5500

CONTENTS

Page

ENVIRONMENTAL HEALTH CRITERIA FOR BERYLLIUM

1. SUMMARY AND CONCLUSIONS 11

   1.1 Identity, physical and chemical properties,
       analytical methods 11
   1.2 Sources of human and environmental exposure 12
   1.3 Environmental transport, distribution, and
       transformation 13
   1.4 Environmental levels and human exposure 13
   1.5 Kinetics and metabolism 14
   1.6 Effects on organisms in the environment 15
   1.7 Effects on experimental animals and *in vitro*
       test systems 17
   1.8 Effects on human beings 18
   1.9 Evaluation of human health risks and effects on the
       environment 21
       1.9.1 Human health risks 21
       1.9.2 Effects on the environment 22

2. IDENTITY, PHYSICAL AND CHEMICAL PROPERTIES,
   ANALYTICAL METHODS 23

   2.1 Identity 23
       2.1.1 Pure beryllium and beryllium compounds 23
       2.1.2 Impure beryllium compounds 23
   2.2 Physical and chemical properties 23
   2.3 Analytical methods 31
       2.3.1 Sampling procedure and sample preparation . . . 31
             2.3.1.1 Sampling 31
             2.3.1.2 Sample decomposition 36
             2.3.1.3 Separation and concentration 36
       2.3.2 Detection and measurement 37

		Page
3.	SOURCES OF HUMAN AND ENVIRONMENTAL EXPOSURE	.40

    3.1 Natural occurrence 40
    3.2 Man-made sources 41
        3.2.1 Industrial production and processing 41
            3.2.1.1 Production levels 41
            3.2.1.2 Manufacturing process 43
            3.2.1.3 Emissions during production and use ..44
            3.2.1.4 Disposal of wastes 46
        3.2.2 Coal and oil combustion 47
    3.3 Uses 47

4. ENVIRONMENTAL TRANSPORT, DISTRIBUTION, AND TRANSFORMATION 54

5. ENVIRONMENTAL LEVELS AND HUMAN EXPOSURE 54

    5.1 Environmental levels 56
        5.1.1 Ambient air 56
        5.1.2 Surface waters and sediments 58
        5.1.3 Soil 60
        5.1.4 Food and drinking-water 61
        5.1.5 Tobacco 62
        5.1.6 Environmental organisms 62
            5.1.6.1 Plants 62
            5.1.6.2 Animals 63
    5.2 General population exposure 64
    5.3 Occupational exposure 66
        5.3.1 Exposure levels 66
        5.3.2 Occupational exposure standards 71
        5.3.3 Biological monitoring 72

6. KINETICS AND METABOLISM 74

    6.1 Absorption 74
        6.1.1 Respiratory absorption 74
        6.1.2 Dermal absorption 76

	Page
6.1.3 Gastrointestinal absorption	.77
6.2 Distribution and retention	.77
6.3 Elimination and excretion	.80
6.4 Biological half-life	.82

7. EFFECTS ON ORGANISMS IN THE ENVIRONMENT ..84

    7.1 Microorganisms .84
    7.2 Aquatic organisms .85
        7.2.1 Plants .85
        7.2.2 Animals .86
    7.3 Terrestrial organisms .86
        7.3.1 Plants .86
        7.3.2 Animals .91

8. EFFECTS ON EXPERIMENTAL ANIMALS AND *IN VITRO* TEST SYSTEMS .92

    8.1 Single exposures .92
    8.2 Short- and long-term exposures .95
        8.2.1 Short-term exposure .95
            8.2.1.1 Oral .95
            8.2.1.2 Inhalation .95
            8.2.1.3 Other .97
        8.2.2 Long-term exposure .97
            8.2.2.1 Oral .97
            8.2.2.2 Inhalation .97
    8.3 Skin irritation and sensitization .98
    8.4 Reproduction, embryotoxicity, and teratogenicity 101
    8.5 Mutagenicity and related end-points .102
        8.5.1 DNA damage .102
        8.5.2 Mutation .102
            8.5.2.1 Bacteria and yeast .102
            8.5.2.2 Cultured mammalian cells .103
        8.5.3 Chromosomal effects .103
    8.6 Carcinogenicity .104
        8.6.1 Bone cancer .104
        8.6.2 Lung cancer .105

	Page
8.7 Mechanisms of toxicity, mode of action	114
8.7.1 Effects on enzymes and proteins	114
8.7.2 Immunological reactions	117

9. EFFECTS ON HUMAN BEINGS 118

9.1 General population exposure 118
9.2 Occupational exposure 120
    9.2.1 Effects of short- and long-term exposure 120
        9.2.1.1 Acute disease 120
        9.2.1.2 Chronic disease 125
9.3 Carcinogenicity 136
    9.3.1 Epidemiological studies 136

10. EVALUATION OF HUMAN HEALTH RISKS AND EFFECTS ON THE ENVIRONMENT 145

10.1 Evaluation of human health risks 145
10.2 Evaluation of effects on the environment 148
10.3 Conclusions 149
    10.3.1 Acute beryllium disease 149
    10.3.2 Chronic beryllium disease 150
    10.3.3 Cancer 150

11. RECOMMENDATIONS 132

12. PREVIOUS EVALUATIONS BY INTERNATIONAL BODIES 153

REFERENCES 155

RESUME 183

RESUMEN 197

WHO TASK GROUP ON ENVIRONMENTAL HEALTH CRITERIA FOR BERYLLIUM

Members

Dr V. Bencko, Institute of Tropical Health, Postgraduate School of Medicine and Pharmacy, Prague, Czechoslovakia

Dr A.W. Choudhry, Division of Environmental and Occupational Health, Kenya Medical Research Centre (KEMRI), Nairobi, Kenya (*Chairman*)

Dr R. Hertel, Fraunhofer Institute of Toxicology and Aerosol Research, Hanover, Federal Republic of Germany

Dr P.F. Infante, Office of Standards Review, Occupational Safety and Health Administration, US Department of Labor, Washington, DC, USA

Professor A. Massoud, Department of Community, Environmental and Occupational Medicine, Ain Shams University, Abbassia, Cairo, Egypt

Dr L.A. Naumova, Institute of Industrial Hygiene and Occupational Diseases, Moscow, USSR (*Vice-Chairman*)

Professor A.L. Reeves, Faculty of Allied Health Professions, Department of Occupational and Environmental Health, Wayne State University, Detroit, Michigan, USA

Dr G. Rosner, Fraunhofer Institute of Toxicology and Aerosol Research, Hanover, Federal Republic of Germany (*Rapporteur*)

Representatives of Nongovernmental Organizations

Dr A.V. Roscin, representative of the International Commission on Occupational Health (ICOH), also a designated national observer, Central Institute for Advanced Medical Training, Moscow, USSR

Observers

    Dr N.A. Khelkovsky-Sergeev, Institute of Industrial Hygiene and Occupational Diseases, Moscow, USSR

Secretariat

    Dr Z. Grigorevskaya, Centre for International Projects, Moscow, USSR (*Project Officer*)

    Dr E.M. Smith, International Programme on Chemical Safety, Division of Environmental Health, World Health Organization, Geneva, Switzerland (*Secretary*)

    Dr V. Turosov (also representing International Agency for Research on Cancer), Cancer Research Center, Academy of Medical Sciences of the USSR, Moscow, USSR

NOTE TO READERS OF THE CRITERIA DOCUMENTS

Every effort has been made to present information in the criteria documents as accurately as possible without unduly delaying their publication. In the interest of all users of the environmental health criteria documents, readers are kindly requested to communicate any errors that may have occurred to the Manager of the International Programme on Chemical Safety, World Health Organization, Geneva, Switzerland, in order that they may be included in corrigenda, which will appear in subsequent volumes.

* * *

A detailed data profile and a legal file can be obtained from the International Register of Potentially Toxic Chemicals, Palais des Nations, 1211 Geneva 10, Switzerland (Telephone no. 7988400/7985850).

ENVIRONMENTAL HEALTH CRITERIA FOR BERYLLIUM

A WHO Task Group on Environmental Health Criteria for Beryllium met at the Ukrania Hotel, Moscow, USSR, from 3 to 7 July 1989, under the auspices of the USSR State Committee for Environmental Protection, Centre for International Projects. Dr S.N. Morozov welcomed the participants on behalf of the host institution and Dr E. Smith opened the meeting on behalf of the three cooperating organizations of the IPCS (ILO/UNEP/WHO). The Task Group reviewed and revised the draft criteria document and made an evaluation of the health risks of exposure to beryllium.

The first draft of this document was prepared by Dr R. HERTEL and Dr G. ROSNER, Fraunhofer Institute for Toxicology and Aerosol Research, Hanover, Federal Republic of Germany. This draft was reviewed in the light of international comments by a Working Group comprising Dr V. BENCKO, Prague, Czechoslovakia, Dr M. PISCATOR, Stockholm, Sweden, Dr F.W. SUNDERMAN, Farmington, Connecticut, USA, with the assistance of Dr R. Hertel and Dr G. Rosner. The revised draft resulting from this Working Group was submitted for the Task Group review. Dr E. SMITH, IPCS Central Unit, was responsible for the overall scientific content of the document and the organization of the meetings, and Mrs M.O. HEAD of Oxford, England, was responsible for the editing.

The efforts of all who helped in the preparation and finalization of the document are gratefully acknowledged.

* * *

Financial support for the Task Group was provided by the United Nations Environment Programme, through the USSR Commission for UNEP. Partial financial support for the publication of this criteria document was kindly provided by the United States Department of Health and Human Services, through a contract from the National Institute of Environmental Health Sciences, Research Triangle Park, North Carolina, USA – a WHO Collaborating Centre for Environmental Health Effects.

1. SUMMARY AND CONCLUSIONS

1.1 Identity, physical and chemical properties, analytical methods

Beryllium is a steel-grey, brittle metal, existing naturally only as the ^{9}Be isotope. Its compounds are divalent. Beryllium has several unique properties. It is the lightest of all solid and chemically-stable substances, with an unusually high melting point, specific heat, heat of fusion, and strength-to-weight ratio. It has excellent electrical and thermal conductivities. Because of its low atomic number, beryllium is very permeable to X-rays. Its nuclear properties include the breaking, scattering, and reflecting of neutrons, as well as the emission of neutrons on α-bombardment.

Beryllium has a number of chemical properties in common with aluminium, particularly its high affinity for oxygen. Hence, a very stable surface film of beryllium oxide (BeO) is formed on the surface of metallic beryllium and beryllium alloys, providing high resistance to corrosion, water, and cold oxidizing acids. When ignited in oxygen, beryllium powder burns with a temperature of 4500 °C. Sintered beryllium oxide ("beryllia") is very stable and possesses ceramic properties. Cationic beryllium salts are hydrolysed in water and react to form insoluble hydroxides or hydrated complexes at pH values in the range of 5–8, and beryllates, above pH 8.

As an additive in alloys, beryllium confers a combination of outstanding properties on other metals, particularly, resistance to corrosion, high modulus of elasticity, non-magnetic and non-sparking characteristics, increased electrical and thermal conductivities, and a greater strength than that of steel.

A variety of analytical methods have been used to determine beryllium in various media. Older methods include spectroscopic, fluorometric, and spectrophotometric techniques. Flameless atomic absorption spectrometry and gas chromatography are the methods of choice; the detection limits are 0.5 ng/sample (flameless atomic absorption) and 0.04 pg/sample (gas chromatography with electron-capture detection). In addition, inductively coupled plasma atomic emission spectrometry is being increasingly used.

1.2 Sources of human and environmental exposure

Beryllium is the 35th most abundant element in the earth's crust, with an average content of about 6 mg/kg. Apart from the gemstones, emerald (chromium-containing beryl) and aquamarine (iron-containing beryl), only 2 beryllium minerals are of economic significance. Beryl contains up to 4% of beryllium and is mined in Argentina, Brazil, China, India, Portugal, the USSR, and in several countries in southern and central Africa. Although it contains less than 1% beryllium, bertrandite has become the main source of this metal in the USA.

The annual global production of beryllium minerals in the period 1980–84 was estimated to be around 10 000 tonnes, which corresponds to approximately 400 tonnes of beryllium. Despite the considerable fluctuations in beryllium supply and demand resulting from sporadic government programmes in armaments, nuclear energy, and aerospace, demand for beryllium was expected, in 1986, to increase at an average annual rate of about 4% up to 1990.

In general, beryllium emissions during production and use are of minor importance compared with emissions that occur during the combustion of coal and fuel oil, which have natural average contents of 1.8–2.2 mg Be/kg dry weight, and up to 100 μg Be/litre, respectively. Beryllium emission from the combustion of fossil fuels amounted to approximately 93% of the total beryllium emission in the USA, one of the main producer countries. Improved control measures can substantially reduce the emission of beryllium from power plants.

Though the combustion of fossil fuels determines the beryllium background levels in ambient air, production-related sources can lead to locally elevated ambient concentrations, particularly where there are insufficient control measures. Similarly, emissions arising from the testing and use of beryllium-powered rockets could be of potential local significance. In occupational settings, exposure occurs mainly during the processing of beryllium ores, metallic beryllium, beryllium-containing alloys, and beryllium oxide.

Summary and conclusions

Production industries exist only in Japan, the USA, and the USSR. In other countries, the imported pure metal, alloys, or ceramic beryllium oxide are processed to end products.

Most beryllium waste results from pollution control measures and is either recycled or buried. Recycling of the majority of end-products is not economically worth while because of their small volume and low beryllium content.

Approximately 72% of the world production of beryllium is used in the form of beryllium-copper and other alloys in the aerospace, electronics, and mechanical industries. About 20% is used as the free metal, mainly in the aerospace, weapons, and nuclear industries. The remainder is used as beryllium oxide for ceramic applications, principally in electronics and microelectronics.

1.3 Environmental transport, distribution, and transformation

Data concerning the fate of beryllium in the environment are limited. Atmospheric beryllium oxide particles return to earth through wet and dry deposition. Within the environmental pH range of 4–8, beryllium is strongly absorbed by finely-dispersed sedimentary minerals, thus preventing release to ground water.

Beryllium is believed not to biomagnify to any extent within food chains. Most plants take up beryllium from the soil in small amounts, and very little is translocated from the roots to other plant parts.

1.4 Environmental levels and human exposure

Beryllium concentrations in surface and drinking-waters are usually in the low µg/litre range. Levels in soils range between 1 and 7 mg/kg. Terrestrial plants generally contain less than 1 mg beryllium/kg dry weight. Amounts of up to approximately 100 µg/kg fresh weight have been found in various marine organisms.

Atmospheric beryllium concentrations at rural sites in the USA ranged from 0.03 to 0.06 ng/m^3. In countries with less fossil fuel combustion, background levels should be lower. Annual average beryllium concentrations in urban air in the USA were found to range from <0.1 to 6.7 ng/m^3. In Japanese cities, an average of

0.04 ng/m^3 was found with the highest values (0.2 ng/m^3) occurring in industrial areas.

Before the establishment of control measures in the 1950s, atmospheric beryllium concentrations were extremely high in the vicinity of production and processing plants. In addition, "paraoccupational" exposure used to occur in workers' families, known as neighbourhood cases, which were related to contact with the worker's clothes, atmospheric exposure, or both. Today, these sources of exposure are normally insignificant for the general population. The principal source of environmental exposure of the general population to airborne beryllium is the combustion of fossil fuels. Exceptionally high exposure could occur in the vicinity of power plants that burn coal containing high levels of beryllium and do not apply adequate control measures. Tobacco smoking is probably another important source of beryllium exposure.

The growing use of beryllium in base dental casting alloys could be of some significance for the general population, because of the high potential of beryllium to provoke contact allergic reactions.

Prior to 1950, exposure to beryllium in working environments was usually very high, and concentrations exceeding 1 mg/m^3 were not unusual. Control measures to meet the occupational standards of 1–5 μg Be/m^3 (time-weighted average), established by various countries, have drastically reduced work-place concentrations of beryllium, though these values are not being achieved everywhere.

Levels of beryllium in tissues or body fluids may be indicative of a previous exposure situation. In persons who have not been specifically exposed, levels in the urine are around 1 μg/litre and those in lung tissue, less than 20 μg/kg (dry weight). The limited data available do not allow the substantiation of a clear relationship between exposure and body burden, though clearly elevated levels (>20 μg/kg) have been found in lung tissue samples from patients with beryllium disease.

1.5 Kinetics and metabolism

There are no human data on the deposition or absorption of inhaled beryllium. Animal studies have shown that, after being deposited in the lungs, beryllium remains there and is slowly absorbed into the blood. Pulmonary clearance is biphasic, with a fast elimination phase in the first 1–2 weeks following cessation of exposure.

Most of the beryllium circulating in the blood is transported in the form of a colloidal phosphate. A significant part of the inhaled dose is incorporated into the skeleton, which is the ultimate site of beryllium storage. Generally, inhalation exposure also results in long-term storage of appreciable amounts of beryllium in lung tissue, particularly in pulmonary lymph nodes. More soluble beryllium compounds are also translocated to the liver, abdominal lymph nodes, spleen, heart, muscle, skin, and kidney.

Following oral administration of beryllium, a small amount (less than 1% of the dose) is generally absorbed into the blood and stored in the skeleton. Small amounts have also been found in the gastrointestinal tract and in the liver.

The absorption of beryllium through intact skin is negligible, as beryllium is bound by epidermal constituents.

A considerable proportion of absorbed beryllium is rapidly eliminated, mainly in the urine, and, to a small extent, in the faeces. Part of inhaled beryllium is also eliminated in the faeces, probably as a result of clearance from the respiratory tract and ingestion of swallowed beryllium.

Because of the long storage of beryllium in the skeleton and lungs, its biological half-life is extremely long. A half-life of 450 days has been calculated for the human skeleton.

1.6 Effects on organisms in the environment

Soil microorganisms grown in a magnesium-deficient medium grow better in the presence of beryllium, because of the partial substitution of beryllium for magnesium in the organisms' metabolism. Similar growth-stimulating effects have been noted in algae and crop plants. This phenomenon seems to be pH-dependent, as it only occurs at high pH. At pH 7 or below, beryllium is toxic for aquatic and terrestrial plants, regardless of the magnesium levels in the growth medium.

Generally, plant growth is inhibited by soluble beryllium compounds at mg/litre concentrations. For example, in bush beans

(*Phaseolus vulgaris*) grown in nutrient solution at pH 5.3, an 88% yield reduction was observed at a concentration of 5 mg Be/litre. Effects were first observed on the roots, which turned brown and failed to resume normal elongation. Roots accumulate most of the beryllium taken up, and very little is translocated to the upper parts of the plant. The critical contents of beryllium resulting in a 50% decrease in yield were estimated to be about 3000 mg Be/kg dry weight and 6 mg Be/kg, respectively, in the roots and outer leaves of cabbage plants (*Brassica oleracea*).

Stunting of both roots and foliage was noted in soil cultures of beans, wheat, and ladino clover, but no chlorosis or mottling of the foliage occurred.

In soil culture, beryllium phytotoxicity is governed by the nature of soil, particularly its cation exchange capacity and the pH of the soil solution. Apart from the magnesium-substituting effect, the diminished phytotoxicity under alkaline conditions also results from the precipitation of beryllium as unavailable phosphate salt.

The mechanism underlying the phytotoxicity of beryllium is probably based on the inhibition of specific enzymes, particularly plant phosphatases. Beryllium also inhibits uptake of essential mineral ions.

In acute toxicity studies on different freshwater fish species, LC_{50} values were found to vary from 0.15 to 32 mg Be/litre, depending on species and test conditions. Toxicity to fish increased with decreasing water hardness; beryllium sulfate was one to two orders of magnitude more toxic for fathead minnows and bluegills in soft water than in hard water. Salamander larvae and the waterflea (*Daphnia magna*) showed a similar sensitivity.

There are no validated data on the long-term toxicity of beryllium in aquatic animals. However, one unpublished study has provided evidence of *Daphnia magna* being adversely affected at considerably lower concentrations (5 μg Be/litre) in long-term reproduction tests than in acute toxicity tests (EC_{50}, 2500 μg Be/litre).

Summary and conclusions

1.7 Effects on experimental animals and *in vitro* test systems

Symptoms of acute beryllium poisoning in experimental animals were respiratory disorders, spasms, hypoglycaemic shock, and respiratory paralysis.

Implantation of beryllium compounds and metallic beryllium in the subcutaneous tissues may produce granulomas, similar to those observed in human beings. Guinea-pigs developed cutaneous hypersensitivity on intradermal injection of soluble beryllium compounds.

As a secondary effect, beryllium carbonate produced rickets (rachitis) in young rats, through intestinal precipitation of beryllium phosphate and concomitant phosphorus deprivation.

Acute chemical pneumonitis occurred in various animal species following the inhalation of beryllium metal or different beryllium compounds, including insoluble forms. Repeated daily exposure to beryllium sulfate mist, at a mean concentration of 2 mg Be/m^3, was lethal for rats (90% deaths), dogs (80%), cats (80%), rabbits (10%), guinea-pigs (60%), monkeys (100%), goats (100%), hamsters (50%), and mice (10%). Because of a synergistic effect of the fluoride ion, the effects of beryllium fluoride were about twice as great as those of the sulfate. Some of the lesions in the lungs resembled those in man, but the granulomas were not identical.

The inhalation toxicity of insoluble beryllium oxide depends to a great extent on its physical and chemical properties, which can alter considerably, depending on production conditions. Because its ultimate particle size is smaller and there is less aggregation, low-fired BeO (400 °C) at 3.6 mg Be/m^3 for 40 days caused mortality in rats and marked lung damage in dogs, whereas high-fired BeO grades (1350 °C and 1150 °C) did not produce pulmonary damage, in spite of a higher total exposure (32 mg Be/m^3, 360 h).

The characteristic non-malignant response to long-term, low-level inhalation exposure to soluble and insoluble beryllium compounds is chronic pneumonitis associated with granulomas, which only partly corresponds to the chronic disease seen in humans beings.

The results of genotoxicity tests indicate that beryllium interacts with DNA and causes gene mutations, chromosomal aberrations, and sister chromatid exchange in cultured mammalian somatic cells, though it was not mutagenic in bacterial test systems.

Intravenous (3.7–700 mg Be) and intramedullary (0.144–216 mg Be) injection of beryllium metal and various compounds produced osteosarcomas and chondrosarcomas in rabbits, with metastases occurring in 40–100% of the animals, most frequently in the lungs.

In rats, inhalation (0.8–9000 µg Be/m^3) or intratracheal (0.3–9 mg Be) exposure to soluble and insoluble beryllium compounds, beryllium metal, and various beryllium alloys induced lung tumours of the adenoma or adenocarcinoma type, partly metastasizing. Beryl (620 µg Be/m^3) was the only beryllium ore that caused lung carcinomas (bertrandite, at 210 µg Be/m^3, did not). Beryllium oxide proved carcinogenic to rats, but the incidence of pulmonary adenocarcinomas was much higher after intratracheal administration (9 mg Be) of a low-fired specification (51%) compared with high-fired oxides (11–16%). At the time of many of these studies, study design and laboratory practice did not usually comply with current practices. Thus, the reported inhalation exposure data should be considered with particular care.

The induction of pulmonary cancer by beryllium is highly species-specific. While rats and, perhaps, monkeys are very susceptible in this respect, no pulmonary tumours have been observed in rabbits, hamsters, or guinea-pigs.

Mechanisms for beryllium toxicity have been based on 3 theories: (1) beryllium affects phosphate metabolism by inhibiting crucial enzymes, particularly alkaline phosphatase; (2) beryllium inhibits replication and cell proliferation by affecting enzymes of nucleic acid metabolism; and (3) beryllium toxicity involves an immunological mechanism, as shown in guinea-pigs, which develop cell-mediated hypersensivity in the skin.

1.8 Effects on human beings

Toxicologically relevant exposure to beryllium is almost exclusively confined to the work-place. Before the introduction of improved emission control and hygiene measures in beryllium plants, several "neighbourhood" cases of chronic beryllium disease were reported. By 1966, a total of 60 cases had been reported in the USA, some of

Summary and conclusions

which were related to contact with workers' clothes ("para-occupational" exposure) or to air exposure in the close vicinity of beryllium plants. No cases have been reported in recent years.

Recently, several cases of an allergic contact stomatitis, probably caused by beryllium-containing dental prostheses, have been reported.

In the 1930s and 1940s, several hundred cases of acute beryllium disease occurred, particularly in workers in beryllium-extraction plants in Germany, Italy, the USA, and the USSR. Inhalation of soluble beryllium salts, particularly the fluoride and sulfate, at concentrations exceeding 100 μg Be/m^3, consistently produced acute symptoms among almost all exposed workers, while, at a level of 15 μg/m^3 and below (determined using out-of-date analytical methods), no cases were registered. After adoption of a maximum exposure concentration of 25 μg/m^3 in the early 1950s, cases of acute beryllium disease drastically decreased.

Signs and symptoms of acute beryllium disease range from mild inflammation of the nasal mucous membranes and pharynx to tracheo-bronchitis and severe chemical pneumonitis. In severe cases, patients died of acute pneumonitis, but in most cases, after cessation of exposure, complete recovery occurred within 1-4 weeks. In a few cases, chronic beryllium disease developed years after recovery from the acute form.

Direct contact with soluble beryllium compounds causes contact dermatitis and possibly conjunctivitis. Sensitized individuals react much more rapidly and to lower amounts of beryllium. Soluble or insoluble beryllium compounds, introduced in, or beneath, the skin produce chronic ulcerations, with granulomas often appearing after several years.

Chronic beryllium disease differs from the acute form in having a latent period ranging from several weeks up to more than 20 years; it is of long duration and progressive in severity. In the US Beryllium Case Registry (a central file on reported cases of beryllium disease, established in 1952) 888 cases were registered up to 1983. Six hundred and twenty two cases were classified as chronic, of which 557 resulted from occupational exposure, mainly within the fluorescent lamp industry (319 cases) or within beryllium extraction plants (101 cases). After the use of zinc beryllium silicate and beryllium oxide in fluorescent tube phosphors was abandoned in 1949, and an occupational exposure limit (TWA, 2 µg Be/m^3) was adopted, cases of chronic beryllium disease dramatically decreased, but new cases resulting from exposure to an air concentration of around 2 µg/m^3 have been recorded.

The term "chronic beryllium disease" is preferred to the term "berylliosis", because this disease differs from a typical pneumoconiosis. Granulomatous inflammation of the lung, associated with dyspnoea on exertion, cough, chest pain, weight loss, fatigue, and general weakness, is the most typical feature; right heart enlargement with accompanying cardiac failure, hepatomegaly, splenomegaly, cyanosis, and finger clubbing may also occur. Changes in serum proteins and liver function, renal stones, and osteosclerosis have also been found to be associated with chronic beryllium disease. The evolution of chronic beryllium disease is not uniform; in some cases, spontaneous remission for weeks or years is encountered, followed by exacerbations. In the majority of cases, progressive pulmonary disease is seen with an increased risk of death from cardiac or respiratory failure. The reported morbidity rates among beryllium workers vary from 0.3 to 7.5%. In patients with chronic beryllium disease, the mortality rates are as high as 37%.

Macroscopically, the lungs may show diffuse changes, with widespread scattered small nodules and interstitial fibrosis. Microscopically, there are sarcoid-like granulomas with varying amounts of interstitial inflammation, which are usually indistinguishable from those in other granulomatoses, such as sarcoidosis or tuberculosis.

History taking and tissue analysis serve as a valuable basis in the diagnosis of beryllium disease, though the presence of beryllium in biological material does not prove the presence of disease. Patch testing is not recommended, because it is not very reliable and is itself highly sensitizing. The most useful diagnostic aids are the macrophage migration inhibition assay and the lymphocyte-blast transformation test.

These methods of measuring hypersensitivity are based on an immune mechanism that probably underlies chronic beryllium disease and the delayed cutaneous and granulomatous hypersensitivity.

The great variability in latency and the lack of dose-response relationships in chronic beryllium disease may be explained by immunological sensitization. Pregnancy seems to be a precipitating "stress factor", as 66% of 95 females, registered among the fatal cases in the US Beryllium Case Registry, were pregnant.

Sources of exposure for patients with beryllium disease also include beryllium metal alloy production, machining, ceramics production and research, and energy production. The present occupational exposure standards may not exclude the development of chronic beryllium disease in sensitized individuals.

In several epidemiological studies, the carcinogenicity of beryllium has been examined among workers employed in two US beryllium production facilities and among clinical cases in a registry of beryllium-related lung conditions, derived from these facilities and other occupations. The results of these studies have been questioned on the grounds of selection bias, confounding from cigarette smoking, and underestimation of the expected number of lung cancer deaths, since mortality rates for the period 1965–67 had been used to estimate expected mortality for the years 1968–75. While the first two issues are unlikely to have played a major role in the excess lung cancer risk, the data presented in this document have been based on an "adjusted" expected number of lung cancer deaths. Significantly elevated risks of lung cancer were noted in all studies.

1.9 Evaluation of human health risks and effects on the environment

1.9.1 Human health risks

Provided that the control measures in the beryllium industry are adequate, general population exposure today is mainly confined to low levels of airborne beryllium from the combustion of fossil fuels. In exceptional cases, where coal with an unusually high beryllium content is burned, health problems could arise. The use of beryllium for dental prostheses should be reconsidered, because of the high sensitization potential of beryllium.

Cases of acute beryllium disease resulting in nasopharyngitis, bronchitis, and severe chemical pneumonitis have drastically decreased and, today, may only occur as a consequence of failures in control measure systems. Chronic beryllium disease differs from the acute form in having a latent period of several weeks to more than 20 years; it is of long duration and progressive in severity. The lung is mainly affected; granulomatous inflammation, associated with dyspnoea on exertion, cough, chest pain, weight loss, and general weakness, is the typical feature. Effects on other organs may be of a secondary nature, rather than systemic. The great variability in latency and the lack of dose-response may still occur today among sensitized individuals who have experienced exposure to a concentration of around 2 $\mu g/m^3$.

Despite some deficiencies in study design and laboratory practice, the carcinogenic activity of beryllium in different animal species has been confirmed.

Several epidemiological studies have provided evidence of an excess lung cancer risk from occupational exposure to beryllium. Although a number of criticisms have been raised about the interpretation of these results, available data lead to the conclusion that beryllium is the most likely single explanation for the excess lung cancer observed in the exposed workers.

1.9.2 Effects on the environment

Data concerning the fate of beryllium in the environment, including its effects on aquatic and terrestrial organisms, are limited. Beryllium levels in surface waters (μg/litre range) and soils (mg/kg dry weight range) are usually low and probably do not negatively affect the environment.

2. IDENTITY, PHYSICAL AND CHEMICAL PROPERTIES, ANALYTICAL METHODS

The element beryllium (Be) was discovered in 1798 by the French chemist Vauquelin, who prepared the hydroxide of beryllium. The metallic element was first isolated in independent experiments by Wöhler (1828) and Bussy (Anon, 1828). Owing to the sweet taste of its salts, the new element was called glucinium (G) by Bussy. Today this name is still used in the French chemical literature. In 1957, Wöhler's name "beryllium" was officially recognized by IUPAC (Ballance et al., 1978).

2.1 Identity

2.1.1 Pure beryllium and beryllium compounds

Pure beryllium is a steel-grey, brittle metal, the first element of the second group (alkaline earths) and the third element of the first period of the periodic table. Its compounds are divalent.

Synonyms, trade names, and the chemical formulae of pure beryllium and some of its compounds are given in Table 1.

2.1.2 Impure beryllium compounds

Impure beryllium compounds are mainly represented by the beryllium ores bertrandite and beryl, and numerous beryllium alloys, some of which are listed in Table 2.

2.2 Physical and chemical properties

Some chemical and physical data of beryllium and selected beryllium compounds are listed in Table 3.

Elemental beryllium has many unique properties (Krejci & Scheel, 1966; Petzow & Aldinger, 1974; Ballance et al., 1978; Newland, 1982; Reeves, 1986) It is the lightest of all solid and chemically stable substances with an unusually high melting point. It has a very low density and very high specific heat (1970 kJ/(kg·K), 25 $^\circ$C),

Table 1. CAS chemical names and registry numbers, synonyms, trade names and atomic or molecular formulae of pure beryllium and beryllium compounds[a]

CAS chemical name	CAS registry number	Synonyms and trade names	Formula
Beryllium	7440-41-7	Beryllium-9; glucinium; glucinum	Be
Acetic acid, beryllium salt	543-81-7	Beryllium acetate; beryllium acetate normal	$Be(C_2H_3O_2)_2$
Hexakis[acetato-0:0]-oxotetraberyllium	19049-40-2	Beryllium acetate, basic; beryllium oxide acetate	$Be_4O(C_2H_3O_2)_6$
Bis[carbonato-(2-)]dihydroxytriberyllium	66104-24-3	Beryllium carbonate; beryllium carbonate, basic; beryllium oxide carbonate	$(BeCO_3)_2 \cdot Be(OH)_2$
Beryllium chloride	7787-47-5	Beryllium dichloride	$BeCl_2$
Beryllium fluoride	7787-49-7	Beryllium difluoride	BeF_2
Beryllium hydroxide	13327-32-7	Beryllium dihydroxide; beryllium hydrate	$Be(OH)_2$
Beryllium oxide	1304-56-9	Beryllia; beryllium monoxide; Thermalox	BeO
Phosphoric acid, beryllium salt (1:1)	13598-15-7	Beryllium phosphate; beryllium hydrogen phosphate	$BeHPO_4$
Phenakite	13598-00-0	Beryllium silicate; beryllium silicic acid; orthosilicate	Be_2SiO_4

Table 1 (continued)

Sulfuric acid, beryllium salt (1:1)	13510-49-1	Beryllium sulfate	BeSO$_4$
Silicic acid, beryllium zinc salt	39413-47-3	Zinc beryllium silicate	Exact composition unknown or undetermined

[a] Adapted from: IARC (1980).

Table 2. CAS chemical names and registry numbers, synonyms, trade names, beryllium content and molecular formulae of beryllium ores and alloys[a]

CAS chemical name	CAS registry number	Synonyms and trade names	Composition	Formula
Bertrandite [$Be_4(H_2Si_2O_9)$]	12161-82-9	Bertrandite; beryllium silicate hydrate	42.1 % BeO; 50.3 % SiO_2; 7.6 % water	$4BeO \cdot 2SiO_2 \cdot H_2O$
Beryl [$Be_3(AlSi_3O_9)_2$]	1302-52-9	Beryl ore; beryllium aluminium silicate; beryllium alumino-silicate	10-13 % BeO; 16-19 % Al_2O_3; 64-70 % SiO_2 1-2 % alkali metal oxides; 1-2 % iron and other oxides	$3BeO \cdot Al_2O_3 \cdot 6SiO_2$
Aluminum alloy, Al, Be	12770-50-2	Beryllium-aluminium alloy; alumin(i)um-beryllium alloy; Lockalloy	62 % Be; 38 % Al	-
Copper alloy, Cu, Be	11133-98-5	Beryllium-copper alloy; beryllium copper	0.3-2.0 % Be; 96.9-98.3 % Cu; 0.2 % min. Ni and Co; 0.6 % max. Ni, Fe, and Co;	-
Nickel alloy, Ni, Be	37227-61-5	Beryllium nickel alloy; nickel-beryllium alloy	2-3 % Be; up to 4 % other additives rest: Ni	-

[a] Adapted from: IARC (1980).

Table 3. Physical and chemical properties of beryllium and selected beryllium compounds[a]

Chemical name	Atomic/ molecular mass	Melting-point (°C)	Boiling-point (°C)	Density (g/cm³)	Crystal system	Solubility
Beryllium	9.01	1278 ± 5	2970 (5 mm Hg)	1.85 (20 °C)	α-close-packed hexagonal, β-body-centred cubic	insol. cold H_2O; sl. sol. hot H_2O; sol. dil. acids and alkalies
Beryllium acetate	127.10	300 (dec.)	-	-	plates	insol. cold H_2O, ethanol and other common organic solvents; slow hydrolysis in boiling-water
Beryllium chloride	79.92	405	520	1.899 (25 °C)	needles	very sol. H_2O, ethanol and diethyl ether; sl. sol. benzene and chloroform
Beryllium fluoride	47.01	544 (800, subl.)	1160	1.986 (25 °C)	amorphous	very sol. H_2O; sol. H_2SO_4 and ethanol
Beryllium hydroxide	43.03	-	-	1.92	powder or crystals	very sl. sol. H_2O and dil. alkali; sol. hot conc. NaOH and acids

Table 3 (continued)

Chemical name	Atomic/molecular mass	Melting-point (°C)	Boiling-point (°C)	Density (g/cm^3)	Crystal system	Solubility
Beryllium nitrate[b]	133.03	60	-	-	deliquescent crystalline mass	very sol. H_2O and alcohol
Beryllium oxide	25.01	2530 ± 30	3900	3.01	hexagonal	0.2 mg/litre H_2O; sol. conc. H_2SO_4
Beryllium sulfate	105.07	550-600(dec.)	-	2.443	-	insol. cold H_2O; converted to tetrahydrate in hot water
Beryllium sulfate tetrahydrate	177.14	100 (-2H_2O)	400 (-4H_2O)	1.713	tetrahedric crystals	425 g/litre H_2O; insol. ethanol; sl. sol. conc. H_2SO_4

[a] Adapted from: IARC (1980), unless otherwise specified.
[b] Windholz (1976).

conc. = concentrated; dec. = decomposition; dil. = dilute; insol. = insoluble; sl. = slightly; sol. = soluble; subl. = sublimation.

heat of fusion (11.7 kJ/mol), sound conductance (12 600 m/s), and strength-to-weight ratio.

Beryllium is lighter than aluminium, but is more than 40% more rigid than steel. It also has excellent electrical and thermal conductivities. The only marked adverse feature is its relatively high brittleness, which has restricted the use of metallic beryllium to specialized applications.

Beryllium occurs naturally only as the ^{9}Be isotope; 4 unstable isotopes with mass numbers of 6, 7, 8, and 10 have been identified. Because of its low atomic number, beryllium is very permeable to X-rays. The neutron emission upon α-bombardment is the most important of its nuclear physical properties, and beryllium can be used as a neutron source. Moreover, its low neutron absorption properties and its high-scattering cross-section determine its characteristics as a suitable moderator and reflector of structural material in nuclear facilities; while most other metals absorb neutrons from the fission of nuclear fuel, beryllium atoms only reduce the energy of such neutrons and reflect them back into the fission zone.

The chemical properties of beryllium differ considerably from those of the other alkaline earths, but it has a number of chemical properties in common with aluminium (Krejci & Scheel, 1966; Petzow & Aldinger, 1974; Reeves, 1986). Beryllium shows a very high affinity for oxygen; on exposure to air or water vapour, a thin film of beryllium oxide (BeO) forms on the surface of the bare metal, providing the metal with a high resistance to corrosion. Like aluminium, beryllium oxide (BeO) is amphoteric. The very stable surface film also renders the metal resistant to water and cold oxidizing acids. Dichromate in water enhances this resistance by forming a protective film of chromate, similar to that formed on aluminium. In powder form, beryllium is readily oxidized in moist air and burns, because of the high entropy of formation of BeO ($23 \cdot 10^3$ kJ/kg), with a temperature of about 4500 °C, when ignited in oxygen.

Beryllium powder reacts with fluorine at room temperature, and with chlorine, bromine, iodide, sulfur, and the vapour of selenium or tellurium to a significant extent only at elevated temperatures (Petzow & Aldinger, 1974). From about 900 °C, nitrogen and ammonia react violently with beryllium to form beryllium nitride

(Be_3N_2). No reaction takes place with hydrogen, even at high temperatures. Melted beryllium reacts with most oxides, nitrides, sulfides, and carbides. Because of its amphoteric character, beryllium is dissolved by dilute acids and alkalis.

Cationic beryllium salts are hydrolysed in water and react to form insoluble hydroxides or hydrated complexes at pH values between 5 and 8, and beryllates above a pH of 8 (Reeves, 1986).

Beryllium oxide ("beryllia") is a colourless crystalline solid or an amorphous white powder with an extremely high melting point, high thermal conductivity, low thermal expansion, and high electrical resistivity. It can either be moulded or applied as a coating to a metal or other base; through the process of sintering (1480 °C), a hard compact mass with a smooth glassy surface is formed (Krejci & Scheel, 1966). The ceramic properties of sintered beryllium oxide make it suitable for the production or protection of materials used at high temperatures in corrosive environments.

A detailed review of the properties of beryllium compounds is given by Krejci & Scheel (1966).

The use of beryllium in alloys is based on a combination of outstanding properties that are conferred on other metals (Petzow & Aldinger, 1974). Low density combined with strength, high melting point, resistance to oxidation, and a high modulus of elasticity make beryllium alloys suitable as light-weight materials that must withstand high acceleration or centrifugal forces. Their advantages over steel include greater resistance to corrosion, higher electrical and thermal conductivities, greater strength, and non-magnetic and non-sparking characteristics. Magnesium alloys containing 0.1% beryllium have a markedly reduced risk of combustion.

Most metals form very brittle intermetallic compounds with beryllium. This and the low solubility of most elements in solid beryllium are the reasons that beryllium-rich alloys have not played a significant role. The only alloy with a high beryllium content is lockalloy containing 62% beryllium and 38% aluminium. Lockalloy has a high modulus of elasticity and low density with reasonable ductility. Aluminium does not form beryllides. Other alloys contain up to 3% beryllium (Petzow & Aldinger, 1974; IARC, 1980).

Of the intermetallic compounds, the beryllides of niobium, tantalum, titanium, and vanadium are gaining interest in the aerospace industry (Stokinger, 1981). Their properties include high strength at elevated temperature and good thermal conductivity and oxidation resistance, combined with densities that are lower than those of refractory metals and many ceramics. The most adverse feature of beryllides is their limited plastic deformability (Walsh & Rees, 1978).

2.3 Analytical methods

Methods for sampling, sample preparation, and the determination of beryllium have been reviewed by Drury et al. (1978) and Delves (1981). Since a detailed review of all the analytical procedures is beyond the scope of this document, only a brief overview is provided, including a summary of methods for the sampling and determination of beryllium in various matrices (Table 4).

2.3.1 Sampling procedure and sample preparation

2.3.1.1 Sampling

Since most environmental samples contain only trace amounts of beryllium, the proper collection and treatment of samples, before analysis, is essential.

Beryllium in air is sampled by means of high-volume samplers using low-ash cellulose fibre, cellulose ester, or fibreglass papers as filters for non-volatile contaminants, and liquid- or solid-filled scrubbers or cold traps to collect volatile forms of beryllium. In the USSR, air sampling is performed on filters made of polyvinyl-chloride fibres plunged in filter-supporters (Izmerov, 1985).

Water samples should be collected in borosilicate glass or plastic containers. It is important to adjust the pH to 5, or below, to prevent losses due to adsorption of beryllium on the surface of containers. Particulate matter should be filtered out and analysed separately.

Table 4. Analytical methods for beryllium and beryllium compounds

Medium	Sampling method	Analytical method	Detection limit	Comments	Reference
Air					
Air	Collect particulates with glass fibre filter; digest ashed filter in boiling HCl and HNO_3 refluxing for 3 h; add EDTA-buffer solution and NaOH to pH 5.5–6.0; add benzene solution of trifluoroacetylacetone; decant chelate; wash with NaOH	Gas chromatography with electron-capture detection	0.04 pg/sample	No interference with several other metals; relatively rapid (40 min); relatively inexpensive chromatograph; applicable for routine analysis of ultratrace concentrations (0.1 ng/m^3)	Ross & Sievers (1972)
Air	Ash glass fibre filter strips; add $HNO_3/HClO_4$ containing indium and yttrium; reflux; concentrate; add HNO_3; centrifuge; add $LiCl_2$ solution	Optical emission spectrometry	5.3 µg/ml		Scott et al. (1976)
Air	Dissolve glass fibre filter in hydrofluoric acid; add HNO_3; boil; dilute	Atomic absorption spectrophotometry Graphite furnace atomic absorption spectrophotometry	2.5 ng/m^3 0.05 ng/m^3		Zdrojewski et al. (1976)

Table 4 (continued)

Air	Extract filter with H_2SO_4; add chrome Azurol S, gum arabic solution and EDTA; adjust to pH 2; add cetylpyridinium bromide and hexamine solution; adjust to pH 5	Spectrophotometry (605 nm)	1 ng/ml sample	Inexpensive	Mulwani & Sathe (1977)
Air	Collect particles on filter; no sample preparation required	Emission spectroscopy using laser-induced breakdown spark	3.6 ng/filter (32 mm diameter, for particles 0.5–5 μm in diameter)	Rapid, near real-time method (3–5 min); due to particle size dependence and interferences, only semi-quantitative	Cremers & Radziemski (1985)
Air	Pass air through mixed cellulose ester membrane filter via personal sampling pump; digest filter in HNO_3/H_2SO_4; evaporate to dryness; dissolve in 2%-NaOH/ 3%-H_2SO_4	Graphite-furnace atomic absorption spectrophotometry	5 ng/sample	Identical with official NIOSH method; suitable for working range 0.5–10 μg/m^3 for a 90-litre air sample	IARC (1986)
Biological samples					
Urine	Add HNO_3 and heat to dryness; concentrate by adding NH_4OH to pH 1.5, EDTA, acetylacetone, NH_4OH to pH 7; separate in funnel; centrifuge; draw off lower layer	Atomic absorption spectrometry, nitrous oxide-acetylene flame	2 μg/litre		Bokowski (1968)

Table 4 (continued)

Medium	Sampling method	Analytical method	Detection limit	Comments	Reference
Organic materials (blood, tissue, food, sewage, mud, etc.)	Digest sample by low temperature ashing or pressure decomposition with HNO_3/HF in teflon tube; eliminate interfering elements with EDTA; add trifluoroacetylacetone in benzene; concentrate by evaporation	Gas chromatography with electron-capture detection	10 pg/sample ($\leq$ 1g)	Suitable for trace levels in limited amounts of organic materials	Kaiser et al. (1972)
Organic materials (blood, muscle, urine, etc.)	Digest sample in teflon tube under pressure; add EDTA - and acetyl acetone; separate Be-complex by liquid-liquid extraction with benzene	Flameless atomic absorption spectrometry; graphite tube treated with $ZrOCl_2$ to increase Be-signal	0.6 µg/litre (for 1-ml urine sample)		Stiefel et al. (1976)
Organic materials	Wet-ash dried tissue in HNO_3/$HClO_4$ in platinum dish; add EDTA and NH_4OH to pH 7–8; add acetylacetone; extract with chloroform; add cyclohexane-diammine-tetraacetic acid and 2-hydroxy-3-napthoic acid reagent	Fluorescence spectrometry	Not specified	Used to determine Standard Reference Materials (SRM)	Wicks & Burke (1977)

Table 4 (continued)

Sample	Preparation	Method	Detection limit	Notes	Reference
Biological tissues (hair, fingernail, faeces)	Digest sample in $HNO_3/HClO_4$; evaporate; dissolve in HNO_3 containing lanthanum	Graphite-furnace atomic absorption spectrometry	1 µg/kg		Hurlbut (1978)
Urine	Add HNO_3 containing lanthanum or add HNO_3 and excess NH_4OH; centrifuge; decant solution; heat	Graphite-furnace atomic absorption spectrometry	0.01 µg/litre	Lanthanum enhanced signal and masked various cations	Hurlbut (1978)
Lung tissue	Digest dried sample in $HNO_3/HClO_4$; heat to dryness and dissolve residue in HNO_3	Graphite-furnace atomic absorption spectrometry	Not specified; reported range: 0.002–0.03 µg/g dry weight	Accuracy proved with reference material and by comparing with other analytical methods	Baumgardt et al. (1986)
Water Fresh water	Acidify with HNO_3 to stabilize solution	Graphite-furnace atomic absorption spectrometry Graphite-furnace atomic emission spectrometry	0.06 µg/litre 2 µg/litre	Used to analyse Standard Reference Material	Epstein et al. (1978)
Sea water	Acidify with HCl; add trifluoroacetylacetone; extract beryllium complex with benzine	Gas chromatography with electron capture detection	18 pg/kg	Precision: 5.5% at 180 pg/kg	Measures & Edmond (1982)

Urine samples are also acidified and can be preserved by adding a 37% formalin solution (Keenan & Holtz, 1964). Precautions must be taken to avoid contamination of the urine sample during collection at the work-place. Animal tissues and vegetable matter can also be preserved by adding formalin, or by cooling. [Formaldehyde is irritant, may cause sensitization, and is a possible human carcinogen.] The use of formalin is discouraged in multielemental analyses, as this preservative contains large amounts of contaminants. Instead, immediate freezing of the samples and storage below -15 °C is recommended (Katz, 1985).

2.3.1.2 Sample decomposition

Organic matter, including air particulate filters, must be destroyed to free the beryllium contents. This is accomplished, by wet digestion using different mixtures of nitric, sulfuric, and perchloric acid, or by dry ashing.

The NIOSH method for the determination of beryllium in air involves the digestion of cellulose ester membrane filters in a mixture of nitric and sulfuric acids (NIOSH, 1984).

Soft tissue and small bone samples can be decomposed by covering the sample with concentrated nitric acid and repeatedly heating to dryness. The dried (400 °C) residue can then be analysed. Large bone samples are dried to constant weight at 105 °C and dry ashed in a muffle furnace by raising the temperature gradually to 500 °C and heating for several hours. The ash residue is extracted with hydrochloric acid (Drury et al., 1978).

Kingston & Jassie (1986) described the use of microwave energy for the acid digestion of organic samples as a time-saving and reliable method.

2.3.1.3 Separation and concentration

Several techniques are used to concentrate or separate beryllium from interfering elements, prior to analysis (Drury et al., 1978). Precipitation of beryllium as the phosphate, hydroxide, or organic complex is only recommended for the separation of macro quantities of beryllium from small amounts of impurities. Alternatively, it can be coprecipitated with calcium, manganese, titanium, and iron phosphates, and with aluminium and iron hydroxides. However, in both cases, considerable losses may occur.

In contrast to precipitation, solvent extraction can be used for micro quantities of beryllium. Organic solvents, such as benzene, chloroform, or carbon tetrachloride, containing a beryllium complexing agent, are added to aqueous solutions in which cationic impurities have been complexed with ethylenediaminetetraacetic acid. The latter does not complex beryllium. After separating the two solvents, the organic phase, which contains the beryllium, is either further processed or directly used in analysis (Drury et al., 1978).

Interfering substances can also be removed by ion exchange techniques, using either cation or anion exchangers, and by electrolysis with a mercury cathode.

2.3.2 Detection and measurement

Older methods used up to the 1960s included spectroscopic, fluorometric, and spectrophotometric techniques. The main deficiency of spectrophotometric methods lies in the non-specificity of the complexing agents used to form coloured complexes with beryllium. The limit of detection with these methods is 100 ng Be/sample (Fishbein, 1984). The fluorometric method, which is based on fluorescent dyes, preferably morin, has a very low limit of detection of 0.02 ng Be/sample; its sensitivity is only exceeded by that of the gas chromatographic method. However, fluorometry may be subject to many errors, unless several time-consuming and cumbersome processing steps are applied prior to analysis. Emission spectroscopy is the most satisfactory method, in terms of specificity and sensitivity. Thermal or electrical excitation of the samples, which must be highly concentrated, is accomplished by the use of a direct or alternating current arc, and alternating current spark, with limits of detection in the range of 0.5–5.0 ng Be/sample (Drury et al., 1978; Fishbein, 1984).

Atomic absorption spectrometry is a rapid and very convenient method for the analysis of environmental samples. The limit of detection for the flame technique is 2–10 ng/ml, or lower when pre-analysis concentration is employed (Bokowski, 1968). The flameless method is much more sensitive. Hurlbut (1978) achieved detection limits of 1 ng/g for faecal, hair, and fingernail samples and of 0.01 ng/ml for urine samples. Addition of lanthanum was found to enhance the absorption signal and eliminate interference by various cations. Stiefel et al. (1980b) provided a detection limit of 0.01 ng/g for immunoelectrophoretic blood fractions. Using graphite-furnace atomic absorption spectrophotometry, the NIOSH procedure for the determination of beryllium in air is recommended for a working range of 0.05–1.0 µg/sample or 0.5–10.0 µg/m^3 of air for a 90-litre sample (NIOSH, 1984).

Inductively coupled plasma atomic emission spectrometry has been introduced to determine beryllium directly in a variety of biological and environmental matrices (Schramel & Li-Qiang, 1982; Wolnik et al., 1984; Awadallah et al., 1986; Caroli et al., 1988). This method is superior to the previous method, because of its high sensitivity and low level of interferences.

Owing to its high sensitivity and specificity, gas chromatography is also used for determining beryllium in environmental and biological media, particularly at ultratrace levels. To convert beryllium into a volatile form, it is commonly chelated with trifluoroacetylacetone and injected into the chromatographic column. Using an electron-capture detector, Taylor & Arnold (1971) determined beryllium in human blood with a detection limit of 0.08 pg Be/sample. When combined with mass spectrometry, sensitivities in the range of 0.04–10 pg Be/sample were achieved (Wolf et al., 1972). Ross & Sievers (1972) developed a routine method for environmental air analysis. At a limit of detection of 0.04 pg Be/sample, beryllium concentrations in the range of 0.49–0.6 ng Be/m^3 could be determined. Because of its carcinogenic properties, a safer alternative to benzene should be considered as a solvent for trifluoroacetylacetone.

In the USSR, a photometric method is used to determine beryllium in the air, with a sensitivity of 0.005 µg/sample (Krivorutchko, 1966). Using gas chromatography, a sensitivity of 1.5×10^{-5} µg/sample is achieved (Yavorovskaya & Grinberg, 1974). By successive treatment of the samples with water, 5% HCl, and fusing with potassium fluoride, the differential determination of water-soluble salts, beryllium metal, and its oxide is possible (Naumova & Grinberg, 1974).

Other analytical techniques, such as polarography, enzyme inhibition, and various types of activation techniques, have been used, but do not play a major role in routine analysis (Drury et al., 1978). Laser ion mass analysis is a promising technique for the identification of beryllium in tissue sections (Jones Williams & Kelland, 1986). Cremers & Radziemski (1985) used the laser-induced spark technique to develop a near real-time method for monitoring airborne beryllium concentrations.

3. SOURCES OF HUMAN AND ENVIRONMENTAL EXPOSURE

3.1 Natural occurrence

Beryllium is the 35th most abundant element in the earth's crust, with an average content of about 6 mg/kg (Mason, 1952). It occurs in rocks and minerals at concentrations of between 0.038 and 11.4 mg/kg (Drury et al., 1978). More than 40 minerals with beryllium as the main constituent are known. Most beryllium minerals were probably formed during the cooling of granitic magmas, which led to an accumulation of crystallization products, usually in association with quartz (Beus, 1966). Thus, the beryllium content generally increases with increasing contents of silica and alkalis. The most highly enriched beryllium deposits are found in granitic pegmatites, in which independent beryllium minerals crystallize (Wedepohl, 1966).

Only two beryllium minerals are of economic significance. Beryl, an aluminosilicate, is mined in Argentina, Brazil, China, India, Portugal, the USSR, and in several countries of southern and central Africa (US Bureau of Mines, 1985a). Beryl contains up to 4% beryllium. In its purest gem quality, it occurs as emerald (chromium-containing beryl), aquamarine (iron-containing beryl), and as some semi-precious stones.

Although bertrandite contains less than 1% beryllium, it became economically important in the late 1960s, because its processing to beryllium hydroxide (section 3.2) is highly efficient. Bertrandite, mined at Spor Mountain, Utah, USA, accounts for about 85% of the US consumption of beryllium ore (US Bureau of Mines, 1982). The total world reserves of beryllium that can be recovered by mining are estimated at 200 000 tonnes (Petzow & Aldinger, 1974).

Clays and residual minerals contain most of the beryllium of the original rocks from which they have been formed by weathering. Clay soils contain between 2 and 5 mg Be/kg, while sandstones contain less than 1 mg/kg and limestones, much less than 1 mg/kg (Griffitts et al., 1977).

The most important source of environmental beryllium is the burning of coal. It can be found in the ash of many coals at concentrations of about 100 mg/kg (Griffitts et al., 1977). Globally, coals contain average concentrations of between 1.8 and 2.2 mg/kg dry weight (US EPA, 1987). Coal samples from Australia, the Federal Republic of Germany, Norway, Poland, the United Kingdom, USA, and USSR showed concentrations of between < 5 and 15 mg Be/kg (Lövblad, 1977). Mineral oils contain up to 100 μg Be/litre (Drury et al., 1978). The occurrence of beryllium in coal and mineral oil is most probably the result of beryllium accumulation in the precursor plants.

The beryllium contents of natural waters and unpolluted air are very low (Fishbein, 1981) (section 5.1.1).

3.2 Man-made sources

3.2.1 Production levels and manufacture

3.2.1.1 Production levels

Beryllium production started in some industrialized countries around 1916 (Petzow & Aldinger, 1974). In the early 1930s, it gained commercial importance following the discovery that beryllium-copper alloys were extraordinarily hard, resistant to corrosion, non-magnetic, did not spark, and withstood high temperatures. In addition, because of its nuclear and thermal properties and high specific modulus, beryllium metal proved attractive for nuclear and aerospace applications, including weapons. This is the main reason that reliable data on the production and consumption of beryllium are scarce and incomplete. Moreover, considerable fluctuations in beryllium supply and demand result from sporadic government programmes in armaments, nuclear energy, and aerospace. For example, the beryllium demand in the USA, created by the programme for the development of the atomic bomb (Manhattan Project), was about equivalent to the total world demand up to 1940 (Newland, 1982).

World production, excluding the USA, parallelled the fluctuations of the beryllium market with 222 tonnes produced in 1965, 320 tonnes in 1969, and 144 tonnes in 1974. Data on US production are now available and the world production of beryllium can be characterized as shown in Table 5. Including its production from bertrandite, the USA appears to be the world's largest producer of beryllium raw materials. Estimated world production of beryllium minerals was between 8873 and 10335 tonnes in the period 1980–84, which corresponds to between 355 and 413 tonnes of beryllium.

Production industries exist only in Japan, the USA, and the USSR. In other countries, the imported pure metal, alloys, or the ceramic beryllium oxide are processed to end-products (Preuss & Oster, 1980).

The doubling of capacity for beryllium-copper strip has been reported by one US producer, to meet the increasing use of this material in electronic devices (US Bureau of Mines, 1985a). Demand for beryllium was expected, in 1986, to increase at an average annual rate of about 4%, up to 1990 (US Bureau of Mines, 1986).

Table 5. World mine production of beryllium (tonnes)[a]

Country	1980	1981	1982	1983	1984
Argentina	1.4	0.3	0.3	1.1	0.7
Brazil (exports)	24.2	37.6	46.8	55.2	55.2
Madagascar	0.4	0.4	0.4	0.4	0.4
Mozambique	0.9	0.8	0.7	0.7	0.7
Portugal	0.8	0.8	0.8	0.8	0.8
Rwanda	4.8	2.6	3.0	1.4	1.6
South Africa, Republic of	-	5.4	2.6	1.0	-
USA	298.0	293.4	218.0	266.6	241.2
USSR	80.0	80.0	80.0	84.0	84.0
Zimbabwe	0.4	1.8	2.3	2.2	2.2
Total	410.9	423.1	354.9	413.4	386.8

[a] Adapted from: US Bureau of Mines (1985a). Data calculated from beryl ore production figures assuming a beryllium content of 4%.

3.2.1.2 Manufacturing process

The first step in the production of pure beryllium metal or beryllium compounds involves the extraction of a concentrate of crystals of beryllium ores by manual selection or, where conditions warrant, by mechanized mining methods (US Bureau of Mines, 1985b).

Two commercial methods are used to process beryl to beryllium hydroxide (Petzow & Aldinger, 1974; Stokinger, 1981; Reeves, 1986). In the fluoride process, beryl is sintered together with sodium silicofluoride, or the less expensive sodium fluoroferrate, at 700–800 °C to convert beryllium oxide to a water-soluble salt (Na_2BeF_4). This is then leached with water and precipitated from the purified solution with caustic soda as beryllium hydroxide.

The sulfate process involves the alkaline or heat processing of beryl and addition of strong sulfuric acid to the fused, quenched, and ground minerals to extract the sulfates of beryllium, aluminium, and other impurities. Following purification of this solution the beryllium sulfate ($BeSO_4$) is precipitated as the hydroxide.

A less complicated procedure has been developed to process the bertrandite ore. The so-called SX-carbonate-process makes caustic pretreatment redundant and involves the direct leaching of beryllium sulfate with sulfuric acid and subsequent precipitation of beryllium hydroxide, which has a comparable high degree of purity (Petzow & Aldinger, 1974).

Beryllium hydroxide is the starting material for the production of beryllium, beryllia, and beryllium alloys. For further processing, it is ignited to form the oxide (BeO) or converted to the fluoride (BeF_2). By means of the thermal reduction of BeF_2 with other metals, mainly magnesium, beryllium metal is obtained, which can be further processed by furnace, by electrolytic refining, or by powder-metallurgical techniques.

The commercial manufacture of copper-beryllium alloys, which are the most important beryllium alloys, involves melting together virgin copper scrap, pure cobalt or a copper-cobalt master alloy, and a copper-beryllium master alloy containing about 4% beryllium (Ballance et al., 1978). The copper-beryllium master alloy is produced by an arc-furnace method in which beryllium oxide is reduced by carbon, in the presence of molten copper.

3.2.1.3 Emissions during production and use

The emissions of atmospheric beryllium in the USA are summarized in Table 6. Natural emissions are negligible compared with man-made emissions. At present, coal combustion in power plants is a main source of beryllium emission. The application of advanced dust emission control techniques could help to cut beryllium emissions considerably. Because of the lack of data, beryllium emissions into the atmosphere, resulting from the military use of beryllium, cannot be accounted for. Although the contribution of metallurgical sources to the overall beryllium pollution is negligible (Table 6), locally elevated ambient concentrations are likely to result from beryllium emissions during production and processing, particularly in case of insufficient control measures. Sources are beryllium extraction plants, ceramic plants, foundries, machine shops, propellant plants, incinerators, rocket-motor test facilities, and open-burning sites for waste disposal (US EPA, 1973).

Beryllium extraction and production plants emit many forms of beryllium including beryl ore dust, beryllium and beryllium oxide acid fume and dust, and a slurry of $Be(OH)_2$ and $(NH_4)_2BeF_4$ (US EPA, 1973). There are no recent data on emissions during the production and processing of beryllium; according to earlier data from the US EPA (1971), about 5 kg of beryllium for every 1000 tonnes of beryllium processed are released into the ambient atmosphere during beryllium production, resulting in a total emission of 6 kg in the USA in the year 1968 (Drury et al., 1978). About 0.45 kg of beryllium in the form of BeO-containing dusts, fumes, and mists are emitted for every tonne processed to beryllia ceramics (US EPA, 1971), amounting to only 14.4 kg of the total US emission from this source in 1968, assuming a production of 32 tonnes of beryllia ceramics (10% of total beryllium demand in 1968). Major emissions result from the cast iron production and fabrication of beryllium alloys and compounds (Table 6). Beryllium emissions into the US atmosphere from production-related sources added up to about 8.6 tonnes in 1968. This is only about 4.4% of the overall emission from all sources.

Table 6. Atmospheric beryllium emissions from different sources in the USA

Source	Total US production (tonnes/year)	Emission factor (g/tonne)	Annual emission (tonnes/year)	Percentage of total emission
Natural sources[a]				
Windblown dust	8.2×10^6	0.6	5	2.48
Volcanic particles	0.41×10^6	0.6	0.2	0.10
Total emission from natural sources			5.2	2.58
Man-made sources				
Beryllium production and processing:				
Mining			negligible[c]	
Ore processing[a]	8×10^3	37.5	0.3	0.15
Be production			negligible[c]	
Ceramic production[c]	32	450	0.014	0.01
Cast iron production[c]	-	-	3.6	1.79
Production of Be alloys and compounds[b]	-	-	5	2.48
Total emission from beryllium production			8.914	4.43
Combustion of fossil fuel:s				
Coal[a]	640×10^6	0.28	180	89.46
Fuel oil[a]	148×10^6	0.048	7.1	3.53
Total emission from combustion of fossil fuels			187.1	92.99
Total beryllium emission from all sources			201.2	100.00

[a] Data from US Environmental Protection Agency (1987).
[b] Data from Drury et al. (1978).
[c] Data from US Environmental Protection Agency (1971).

As outlined in section 3.3, considerable amounts of beryllium are used for military purposes, e.g., as rocket propellant. During test flights, a major part of the beryllium will be released to the atmosphere (Drury et al., 1978).

3.2.1.4 Disposal of wastes

Most of the beryllium scrap is resold to the producer; recycling of most end-products is not worthwhile, because of their small size and, usually, their low beryllium content (Griffitts et al., 1977). They are either discarded along with other solid wastes or salvaged for the copper in the alloy. In one instance, beryllium-copper dust was dumped on to railroad tracks (OSHA, Personal communication, 1989).

The major portion of beryllium waste results from pollution control measures (Powers, 1976). The beryllium-containing dust retained in scrubbers, electric sleeve filters, and multi-staged purification devices is recycled into the production process, as are liquid and solid wastes from hydrometallurgical and other processes (Izmerov, 1985). Waste waters must be filtered before discharging into the receiving waters. High efficiency is achieved in sewage purification using filters made of "lavsan" to remove metallic beryllium particles (Bobrischev-Pushkin et al., 1976).

Liquid, solid, or particulate waste that is too dilute to recycle is buried in water-proof tailings ponds or in plastic containers sealed in metal drums (US EPA, 1973). Often these wastes are first burned to produce the chemically inert beryllium oxide. The exhaust gases are scrubbed to retain particulates. The disposal of scrap beryllium propellant involves underground detonation and subsequent filtering of exhaust gases through particulate air filters.

The flue gas cleaning system of industrial waste incineration plants is designed to meet the national emission standards. For instance, according to the Clean Air Act of the Federal Republic of Germany, the sum of the emissions of beryllium, benz(a)pyrene, and dibenzanthracene must not exceed 0.1 mg/m^3 in the flue gases of hazardous waste incinerators. Test runs showed an average beryllium emission of 0.02 mg/m^3 (STP) (Erbach, 1984).

3.2.2 Coal and oil combustion

A major source of atmospheric beryllium is the combustion of fossil fuels, of which coal is the most important pollutant source. The US EPA (1987) estimated that between 10 and 30% of the beryllium contained in coal is emitted during the combustion process. The remainder is retained by the captured fly ash. On the basis of the average beryllium content of coal (1.4 mg/kg), the combustion of 790×10^6 tonnes of coal in the USA during 1984 resulted in a total beryllium emission of 220 ± 110 tonnes/year, while the consumption of 110×10^6 tonnes of fuel oil led to a beryllium release of more than 7.1 tonnes. In 1981, the total beryllium emission from fossil fuel combustion was 187.1 tonnes/year or about 93% of the combined emissions from all sources (Table 6).

The emission factor is dependent on the efficiency of mechanical and electrostatic precipitors of power plants. Thus, improved dust emission control measures will cut the emission of pollutants substantially. For instance, the average efficiency of fly-ash collectors (electrostatic precipitators) in coal-power plants in the Federal Republic of Germany is assumed to be between 97 and 99%. Hence, only 2.1 tonnes of beryllium were calculated to be released into the atmosphere in 1981 from the combustion of about 82×10^6 tonnes of coal (Brumsack et al., 1984).

3.3 Uses

Some of the most current applications of beryllium are listed in Table 7.

Almost all of the beryllium produced is used as the free metal, in the form of its alloys, or as the oxide. The various beryllium compounds are primarily used as intermediates in the preparation of beryllium metal or its alloys. Zinc beryllium silicate and beryllium oxide were used widely in fluorescent tube phosphors, until this application was abandoned in 1949 because of considerable health hazards (Newland, 1982).

The main use of beryllium arises from the outstanding properties that it confers on other metals; about 72% of the beryllium produced is used in the form of beryllium-copper and other alloys.

Table 7. Uses of beryllium metal, beryllium alloys, and beryllium oxide as related to their properties[a]

Form	Properties	Technology	Use
Beryllium metal	High strength-to-weight ratio	Aerospace	Windshield frames in US space shuttles Structural components in aeroplanes, rockets, satellites, and space vehicles[b] Antennae in data-gathering satellites[c] Turbine rotor blades
	Heat sink		Aircraft brakes Heat shields for space vehicles and missiles
	Dimensional stability		Inertial guidance systems Other control systems Mirror components of satellite optical systems
	High heat-of-combustion-to weight ratio		Rocket propellant
	Neutron source and moderator	Weapons	Nuclear weapons
	Moderator and reflector for neutrons	Nuclear	Components of nuclear reactors Nuclear fuel element as UBe_{13} alloy[d] Neutron reflector in high-flux test reactors
	Transparency for X-rays	X-ray and radiation	Windows in X-ray tubes and radiation detection devices Coating for biological X-ray microanalysis[b]
	Transparency for X-rays	Computer	Ultrathin foil for X-ray lithography
Beryllium-copper alloys	Dimensional stability	Aerospace	Aircraft engine parts

Table 7 (continued)

Form	Properties	Technology	Use
Beryllium-copper alloys	High strength; good electrical and thermal conductivity	Electronic	Contacts Switches Circuit breaker parts Fuse clips High-frequency connector plugs
	High strength	Mechanical	Springs Bearings Gear parts Camera shutters Golf club heads[e]
	Non-sparking		Tools
	High strength; good thermal conductivity; dimensional stability	Others	Injection moulds for plastics Precision castings Diaphragms Welding electrodes
Beryllium-aluminium alloy (Lockalloy)		Aerospace	Construction materials for aircraft and spacecraft[f]
Beryllium-copper-cobalt-alloy[f,g]	High strength Good conductivity	Electronic	Springs[f] Switches[f] Contacts[f] Welding electrodes and holders[g]
		Mechanical	Bushings[g] Bearings[g] Soldering iron tips[g]
		Others	Nozzles for gas and oil burners[g] Plunger tips for die-casting machines[g]
Beryllium-nickel-alloy	Higher thermal conductivity than beryllium-copper alloys	Aerospace	Aircraft and spacecraft parts[h]
		Glass	Various glass-moulding functions[f]

49

Table 7 (continued)

Form	Properties	Technology	Use
Beryllium-nickel alloy		Electronic	Electrical connectors[f]
		Others	Springs[f] Diamond drill bit matrices[h] Watch balance wheels[h]
Beryllium-nickel-chromium alloy	Facilitated castability High porcelain-metal bond strength	Dentistry	Alternatives to gold alloys used for crowns and bridgesi[k]
Beryllium oxide	High thermal conductivity, heat capacity, and electrical resistivity	Aerospace	Rocket-chamber-combustion liners
	Ceramic properties	Electronic	Electrical insulator Resistor cores[h] Integrated circuit chip carriers[h] Radio, laser, and microwave tubes[h] Spark plugs Other high-voltage electrical components
	Moderator and reflector for neutrons	Nuclear	Components of nuclear reactors[j]
	High thermal conductivity	Others	Mantels in gas lanterns[m]

[a] Adapted from: Newland (1982), unless otherwise specified.
[b] Lupton & Aldinger (1983).
[c] Greenfield (1971).
[d] Stokinger (1981).
[e] OSHA (1989).
[f] Ballance et al. (1978).
[g] IARC (1980).
[h] Reeves (1986).
[i] Covington et al. (1985a).
[k] Bencko (1989).
[l] Boland (1958).
[m] Griggs (1973).

Sources of exposure

About 20% is used as the free metal (Reeves, 1986) and beryllium oxide accounts for the remaining 8%. In 1983, it was estimated that about 65% of the US consumption was in the form of beryllium alloys, about 15% in the oxide form, and the remainder in the metal form (US Bureau of Mines, 1985b).

Most beryllium-copper alloys are used in parts that need extraordinary hardness, such as bushings, bearings, springs, electric contacts, and switches. A more recent use is the manufacture of 1.8% beryllium-copper alloy golf club heads. Other important applications include the manufacture of welding electrodes and precision casting for optical and mechanical recording instruments. Non-sparking tools made of beryllium-copper alloys are convenient materials for use in explosive atmospheres, e.g., in petroleum refineries.

Beryllium-aluminium alloys are gaining interest for use as construction materials in aircraft and spacecraft technology (Izmerov, 1985). Also, some intermetallic compounds of beryllium, particularly the beryllides of niobium, tantalum, titanium, vanadium, and the borides of beryllium are being considered for use as structural materials for space vehicles (Stokinger, 1981).

Beryllium-nickel alloy is used in the place of beryllium-copper alloy for high-temperature applications (Farkas, 1977). It also has greater hardness than the copper alloy and is therefore used in the production of components requiring this property, e.g., watch-balance wheels.

Beryllium-containing alloys are increasingly used in dentistry as alternatives to more expensive gold alloys. When added to nickel-chromium alloys, beryllium (2% or less) facilitates castability and increases the porcelain-metal bond strength (Covington et al., 1985a). Apart from dental prostheses, beryllium has also been found in the cement used to fix crowns and bridges, as reported by Schönherr & Pevny (1985), but no data were given.

The applications of beryllium metal are mainly related to its nuclear and thermal properties and high specific modulus (Greenfield, 1971; Ballance et al., 1978). For many years, the major application of beryllium in the form of thin sheets, was in X-ray windows. Such windows are also used in Geiger proportional and scintillation counters.

In the 1950s, beryllium and its oxide were believed to be promising moderator and reflector materials for nuclear reactors. However, because the shut-down of a graphite-moderated reactor is much faster and because of the high cost of beryllium, nuclear applications have been limited to test reactors and, probably, to mobile reactors, as used in nuclear submarines. It is also assumed that beryllium is used as reflector material in atomic bombs. The development of beryllium-based cladding material for uranium dioxide fuel was abandoned in the 1960s, because of the high costs and brittleness of beryllium, which caused tube cracking (Greenfield, 1971; Buresch, 1983).

A major application of beryllium is for aircraft and spacecraft structural materials where a combination of light weight, rigidity, dimensional stability, and good thermal characteristics is demanded. The advantage of using beryllium in missiles and spacecraft lies in its superiority, in this respect, over any other metal or alloy. For instance, basic weights of 3-stage missiles are reduced by 40%, compared with steel (Greenfield, 1971). Its high modulus of elasticity and dimensional stability make beryllium an excellent material for use in aircraft and spacecraft instruments including inertial guidance devices and other control systems using gyroscopes, gimbals, torque tubes, and high-speed rotating elements.

Beryllium is also used as a heat sink for aircraft wheel brakes and the heat shields of re-entry space vehicles and missiles, and for scanning mirrors and large mirror components of satellite optical systems (Ballance et al., 1978).

Since most of these applications of beryllium are for military devices, no data concerning the demand for beryllium for the various applications are available. There is almost a total lack of information on the use of beryllium powder as solid rocket propellant, although test flights of such rockets could contribute considerably to local non-occupational exposure to atmospheric beryllium. It is believed that beryllium is gaining interest as an ideal rocket propellant in terms of high heat of combustion at low weight. In these properties, it is superior to other solid and chemically stable substances (Reeves, 1977; Krampitz, 1980). The existence of a rocket propellant industry (Fishbein, 1981) is indicative of the importance of beryllium for this application.

Owing to its high thermal conductivity and heat capacity combined with high electrical resistivity, beryllium oxide has major ceramic applications in electronics and micro-electronics, where it is used as an electrical insulator in parts requiring thermal dissipation. Metallized beryllia is used for the removal of heat in semiconductor devices and integrated circuits (Walsh & Rees, 1978). Its high transparency to microwaves renders beryllium oxide suitable for use in microwave technology (Ballance et al., 1978). The oxide of beryllium is also a component of mantles of gas lanterns, one of the rare non-industrial applications (Griggs, 1973).

4. ENVIRONMENTAL TRANSPORT, DISTRIBUTION, AND TRANSFORMATION

Data concerning the fate of beryllium in the environment are limited. Since the major source of atmospheric beryllium is coal combustion, the most prevalent chemical form is probably beryllium oxide, mainly bound to particles smaller than 1 μm. The residence time of these particles in the atmosphere is about 10 days (US EPA, 1987). Beryllium returns to earth by wet and dry deposition in a similar manner to other metals and on particles of comparable size distribution (Kwapulinski & Pastuszka, 1983).

During the natural processes of weathering and formation of sediments, beryllium resembles aluminium in that it is enriched in clays, bauxites, recent deep-sea deposits, and other hydrolyzate sediments (Newland, 1982).

Reactions of beryllium in solution and soil depend on the pH. At environmental pH ranges of 4–8, beryllium oxide is highly insoluble, thus preventing mobilization in soil. Beryllium is strongly absorbed by finely dispersed sedimentary materials including clays, iron hydroxides, and organic substances (Izmerov, 1985). Thus, very little is released into ground water, during weathering. If beryllium oxide is converted to the ionized salts (chloride, sulfate, nitrate) during atmospheric transport, solubility upon deposition and, hence, mobility in soils would be greatly enhanced, but this has not been reported in the literature.

Because of the low solubility of beryllium oxide and hydroxide at pH levels commonly found in natural waters, only small amounts of beryllium are found in the form of the chloride, fluoride, chlorocarbonate, or organic complexes (Griffitts et al., 1977).

If beryllium is bioavailable in the soil matrices, it can be assimilated by plants and, thus, enter the food chain. Beryllium is classified as a fast-exchange metal, and could potentially interfere with the transport of nutritive metals, such as calcium, into eukaryotic cells (Wood & Wang, 1983). Although there is a lack of data on beryllium levels in environmental organisms representing high trophic levels, and on the fate of beryllium in ecosystems, it is not believed to biomagnify within food chains to an extent that would imply an important pathway to the consumer.

From the beryllium levels found (section 5.1.6.1), it appears that plants take up beryllium in small amounts. However, some species act as accumulators of beryllium. Hickory trees (*Carya* spp.) contain as much as 1 mg/kg dry weight (Griffitts, 1977). Nikonova (1967) found up to 10 mg/kg in several plant species in the South Urals (USSR). Tundra plants (Seward Peninsula, Alaska) tend to accumulate beryllium from soils, if the soil content is at, or below, 20 mg/kg; however, above 50 mg/kg, the plants cannot absorb more (Sainsbury et al., 1968). Thus, plant ash may contain greater amounts of beryllium than the soil.

Tolle et al. (1983) investigated beryllium accumulation by plants grown on soil that had been treated with beryllium-containing precipitor fly ash from a power plant. The beryllium concentration in the soil was not reported. Beryllium uptake by mixed-species crops of alfalfa, timothy, and oats, planted either in agricultural microcosms or in field plots, did not differ from that of control plants. Moreover, uptake by oat grains was comparable to uptake by oat stalks, indicating that there was not any selective enrichment in the grains. Kloke et al. (1984) estimated that the transfer plant/soil coefficient for beryllium was of the order of magnitude of 0.01–0.1, depending on the plant species and soil properties.

The roots of barley, bean, tomato, and sunflower plants, grown for 30 days in aerated nutrient solution containing undetermined levels of the Be^7 isotope, showed a radioactivity of between 29 717 and 66 968 cpm/g. From the corresponding values in the leaf, stem, and fruit (146–910 cpm/g), it can be concluded that very little beryllium is translocated to other plant parts (Romney & Childress, 1965). Leaves seemed to take up more beryllium than stems or fruits.

As with other metals, beryllium contamination also occurs from the wet and dry deposition of beryllium-containing particles on the above-ground plant parts. Although beryllium levels in the leaves of some species have been reported (section 5.1.6), there are no data concerning the uptake of atmospheric beryllium into leaves.

No data are available on the trophic transfer of beryllium in aquatic ecosystems.

5. ENVIRONMENTAL LEVELS AND HUMAN EXPOSURE

5.1 Environmental levels

5.1.1 Ambient air

The atmospheric background level of beryllium in the USA has previously been reported to average less than 0.1 ng/m^3 (Bowen, 1966) or 0.2 ng/m^3 (Sussman et al., 1959). In a more recent survey, the annual averages during 1977-81, at most monitoring stations throughout the USA, were around the detection limit of 0.03 ng/m^3 (US EPA, 1987). This agrees with the mean values of 0.03-0.06 ng/m^3 found by Ross et al. (1977), at rural sites in the USA, using a sensitive chelation-gas chromatographic method. Since fossil fuel combustion contributes to the ubiquitous occurrence of beryllium, particularly in the highly industrialized northern hemisphere, these background levels reflect the overall pollution from this source.

There appears to be a considerable range of reported values for beryllium concentrations in urban air. The air of over 100 cities in the USA, sampled in 1964-65, did not contain detectable amounts of beryllium at a detection limit of 0.1 ng/m^3 (Drury et al., 1978). In the 1950s, beryllium concentrations of between 0.1 and 0.5 ng/m^3 were found in major US cities, such as New York and Los Angeles. The maximum level of beryllium in the air of more than 30 metropolitan areas was 3 ng/m^3 (Chambers et al., 1955).

The highest 24-h level measured in a 1977 survey in Atlanta, Georgia, was 1.78 ng/m^3. The annual averages, at urban monitoring stations throughout the USA with levels exceeding 0.1 ng/m^3, ranged between 0.1 and 6.7 ng/m^3, during 1981-86 (US EPA, 1987). Ross et al. (1977) reported concentrations of beryllium in air particulates of 0.04-0.07 ng/m^3, at suburban sites, and 0.1-0.2 ng/m^3, at urban industrial sites, in Dayton, Ohio.

Using flameless atomic absorption spectrophotometry, Ikebe et al. (1986) found an average of 0.042 ng/m^3 in 76 air samples from

17 Japanese cities, collected between 1977 and 1980. The highest values were found in Tokyo (0.222 ng/m^3) and in an industrial area in Kitakyushu (0.211 ng/m^3).

Freise & Israel (1987) found annual mean values in Berlin ranging between 0.2 and 0.33 ng/m^3, for sectors with different wind direction.

A concentration of 0.06 ng/m^3 was measured in a residential and office area as well as in the inner city area of Frankfurt (Federal Republic of Germany), whereas a concentration of 0.02 ng/m^3 was measured in a rural area near Frankfurt (Mueller, 1979).

Before the introduction of control measures, atmospheric beryllium concentrations were extremely high in the vicinity of point sources. In the vicinity of a Pennsylvania (USA) processing plant, a mean concentration of 15.5 ng/m^3 and a maximum of 82.7 ng/m^3 were reported. At various distances from the plant, the average concentrations dropped from 28 ng/m^3 at 0–800 m to about 6.6 ng/m^3 at a distance of 1800–3200 m and to 1.4 ng/m^3 further away. During a partial plant shutdown, the beryllium level dropped to 4.7 ng/m^3, while a complete 2-week shutdown resulted in an average of 1.5 ng/m^3 (Sussman et al., 1959).

About 0.4 km from the stack of a beryllium emission source in the USA, the beryllium concentration in air was 200 ng/m^3; however, at a distance of 16 km, it was below the detection limit (1 ng/m^3) (Eisenbud et al., 1949). The air level, 400 m from a beryllium extracting and processing plant in the USSR that was not equipped with emission control devices, averaged 1000 ng Be/m^3; at 1000 m, it was between 10 and 100 ng/m^3. Between 500 and 1500 m from a mechanical beryllium-finishing plant with operational filter facilities, no beryllium was detected in the air (Izmerov, 1985).

Bobrischev-Pushkin et al. (1973, 1976) investigated the atmospheric air conditions around a plant for the mechanical treatment of beryllium. In accordance with health rules for operations involving beryllium, the air emissions were subjected to a multi-stage purification process using PVC fibre tissue as a final filtering element. In 290 air samples taken at distances of 500, 900, 1000, and 1500 metres downwind from the plant, beryllium could not be detected by gas chromatography (sensitivity 1.5×10^{-5} µg in the volume analysed).

Bencko et al. (1980) reported beryllium concentrations of between 3.9 and 16.8 ng/m^3 (average 8.4 ng/m^3) in the vicinity of a Czechoslovakian power plant, situated at the edge of a town from which the non-occupationally exposed group in this study was taken.

5.1.2 Surface waters and sediments

Beryllium concentrations in surface waters are usually in the ng/litre range (Table 8). Levels reported for Australian rivers ranged from not detectable to 0.08 μg/litre, with mean concentrations of between 0.02 and 0.03 μg/litre (Meehan & Smythe, 1967). Although otherwise highly polluted, samples of the rivers Rhine and Main (Federal Republic of Germany) contained beryllium only at concentrations of < 0.005–0.02 μg/litre, with mean values of 0.009 and 0.019 μg/litre, respectively (Reichert, 1974). Higher levels were reported in the Rhine region in 1983-85 (IAWR, 1986): the mean values at two measuring stations were around 0.1 μg/litre, the maximum values were between 0.26 and 0.52 μg/litre. Durum & Haffty (1961) analysed 15 major rivers in the USA and Canada and found detectable amounts of beryllium in only 2 water samples (< 0.06 μg Be/litre and < 0.22 μg Be/litre) out of 59.

Beryllium levels in seawater are ten times lower than those in surface waters. In the Pacific Ocean, concentrations of 0.6 ng/litre (Merril et al., 1960) and 2 ng/litre (Meehan & Smythe, 1967) were reported. Data reported by Measures & Edmond (1982) showed that still lower concentrations can be expected. In a detailed profile analysis, the concentration of beryllium has been shown to increase with depth. The mixed layer, up to about 500 m, is characterized by a level of between 0.04 and 0.06 ng Be/litre; the concentrations rise through the main thermocline to levels of 0.22–0.27 ng/litre (25–30 pmol/kg) in the deep and bottom waters (2500–5900 m).

When several ground-water samples were analysed in the western USA, beryllium was detected in only 3 highly acidic mine waters (Griffitts et al., 1977). Ground-water samples from the Federal Republic of Germany contained levels ranging from not detectable (< 0.005 μg/litre) to 0.009 μg/litre with a mean of 0.008 μg/litre (Reichert, 1974).

Table 8. Beryllium concentrations in surface waters

Number of samples	Surface water and location	mg/litre range	mean	Reference
River				
1	Lachlan (Farbes, Australia)		0.01	Meehan & Smythe (1967)
1	Macquarie (Bathurst, Australia)		0.01	
1	Nepean (Emu Plains, Australia)		ND[a]	
27	Woronara (Discharge Pt, Australia)	0.01-0.012	0.03	
26	Woronara (Tolofin, Australia)	0.01-0.08	0.02	
59	15 rivers in the USA and Canada	ND-<0.22[b]	ns[c]	Durum & Haffty (1961)
ns[c]	Raw surface waters in the USA	0.01-1.22[d]	0.19	National Academy of Sciences (1977)
ns[c]	River Rhine (Federal Republic of Germany)	<0.005-0.011	0.009	Reichert (1974)
ns[c]	River Main (Federal Republic of Germany)	0.008-0.02	0.019	
ns[c]	River Rhine (Netherlands)	0.36-0.58[e]	0.09	IAWR (1986)
Sea				
1	Pacific Ocean	-	0.002	Meehan & Smythe (1967)
1	Indian Ocean	-	0.001	
5	Pacific Ocean	-	0.0006	Merril et al. (1960)
ns[c]	Pacific Ocean, near Hawaii, depth 40 m	-	0.00004	Measures & Edmond (1982)

[a] ND = not detected.
[b] Detected but less than figure indicated.
[c] ns = not specified.
[d] Beryllium was detected only in 5.4% of 1577 raw surface waters.
[e] Range of maximum values.

The beryllium contents of sediments correspond to those of soil samples (section 5.1.3). Bottom sediments of lakes in Illinois, USA, contained 1.4–7.4 mg/kg (Dreher et al., 1977). The mean beryllium content of Tokyo Bay and Sagami Bay sediments (Japan) was 1.29 mg/kg (Asami & Fukazawa, 1985).

5.1.3 Soil

As outlined in section 3.1, beryllium is widely distributed in soils at low concentrations. Geochemical surveys (e.g., US EPA (1987)) suggested an overall average of about 6 mg/kg for beryllium in the lithosphere as a whole. More specific data (Shacklette et al., 1971) indicated lower levels for agricultural soils; 847 samples collected at a depth of 20 cm throughout the USA contained between less than 1 and 7 mg beryllium/kg, averaging 0.6 mg/kg. Only 12% of the samples exceeded 1.5 mg/kg. None were collected in geological areas containing large deposits of beryllium minerals. These areas are relatively rare, but they account for the overall lithospheric average of 6 mg/kg.

The beryllium contents of uncontaminated Japanese soils were of the same order of magnitude. Asami & Fukazawa (1985) analysed over 100 soil horizons from all over Japan and found a mean concentration of 1.31 mg beryllium/kg. The beryllium contents of the surface soils of paddy fields ranged from 1.10 to 1.95 mg/kg and those of the subsoils, 0.88–1.95 mg/kg. Podzol and brown forest soil contained between 0.01 and 2.72 mg/kg, with regional differences. Mineral surface soils showed beryllium levels of 0.27–1.66 mg/kg. Beryllium distribution in the profiles of forest soils reflected a leaching process; in all the profiles, the beryllium contents generally increased with an increase in depth. For example, beryllium contents of a yellowish brown forest soil were as follows: 1.66 mg/kg in the topmost mineral layer (A_1-horizon, 0–9 cm), 2.39 mg/kg in the subsoil (B_2-horizon, 17–30 cm), and 2.72 mg/kg in the layer below (C_3-horizon, 48–58 cm). In some profiles, beryllium contents decreased again at deeper horizons.

In some small and unpopulated areas in which rocks contained unusually high levels of beryllium, the overlying soils also showed relatively high beryllium concentrations. For instance, soils of the beryllium district in the Lost River Valley, Alaska, contained up to

300 mg beryllium/kg, with an average of 60 mg/kg (Shacklette et al., 1971).

In the Federal Republic of Germany, an allowable concentration of 10 mg/kg air-dried soil was proposed by Kloke et al. (1984) as a guideline value for arable soils.

5.1.4 Food and drinking-water

Only limited data concerning the beryllium contents of foods are available. Meehan & Smythe (1967), using a chemical analytical method, found generally low levels in various food samples from New South Wales, Australia including: beans, 10 µg/kg ash weight (0.07 µg/kg fresh weight (FW)); cabbage, 30 µg/kg (0.23 µg/kg FW); eggs, 6 µg/kg (0.05 µg/kg FW); milk, 20 µg/kg (0.17 µg/kg FW); crabs, 100–170 µg/kg (15.4–26.2 µg/kg FW); whole marine fish, 21–23 µg/kg (10.6–10.9 µg/kg FW); fish fillets, not detectable– 40 µg/kg (up to 1.48 µg/kg FW); and oyster flesh, 30–100 µg/kg (0.6–2.0 µg/kg FW).

Food samples from the Federal Republic of Germany, analysed using atomic absorption spectrometry, contained the following beryllium concentrations (Zorn & Diem, 1974): crispbread, 120 µg/kg dry weight; green head lettuce, 330 µg/kg; tomatoes, 240 µg/kg; polished rice, 80 µg/kg; and potatoes, 170 µg/kg. Assuming an average moisture content of about 95% for tomatoes and lettuce, 77% for potatoes, and 17% for rice (Franke, 1985), these values correspond to about 10–70 µg/kg, on a fresh weight basis. Although these levels are somewhat higher than those reported above, no direct comparison can be made, because of insufficient numbers of samples and probable differences in sensitivity of the analytical methods used in the studies.

Awadallah et al. (1986) recently published data on beryllium concentrations in some Egyptian crop plants, measured by inductively coupled plasma-atomic emission spectrometry. The plants contained between 0.3 and 2.5 mg/kg dry weight. However, the values reported are probably from single measurements and, thus, the data base is very limited.

Levels of 0.02–0.17 µg Be/litre with a mean of 0.1 µg Be/litre have been reported for drinking-water in the USA (National Academy of Sciences, 1977). Detectable amounts were found in only 1.1% of the samples. According to APHA (1971), beryllium concentrations in US drinking-water ranged from 0.01 to 0.7 µg/litre, with a mean of 0.013 µg/litre. In drinking-water samples from the Federal Republic of Germany, beryllium concentrations ranged between "not detectable" (< 0.005 µg/litre) and 0.009 µg/litre (mean 0.008 µg/litre) (Reichert, 1974). Sauer & Lieser (1986) found beryllium levels of 0.025 ± 0.013 µg/litre (SD), 0.027 ± 0.008 µg/litre, and 0.024 ± 0.007 µg/litre, respectively, in unfiltered, filtered (0.45 µm filter), and ultrafiltered (2 nm filter) drinking-water samples from Wiesbaden, Federal Republic of Germany. Thus, about 95% of the beryllium is in its molecular-dispersed form.

5.1.5 Tobacco

Zorn & Diem (1974) determined the beryllium contents in 3 brands of cigarettes, using atomic absorption spectrometry. The origin of the tobaccos and the number of samples analysed were not indicated. The beryllium levels were 0.47, 0.68, and 0.74 µg/cigarette. Assuming a tobacco content of about 0.6 g/cigarette, the tobacco contained between 0.8 and 1.2 mg Be/kg. Between 1.6 and 10% of the beryllium content, or 0.011–0.074 µg/cigarette, was reported to pass into the smoke during smoking.

5.1.6 Environmental organisms

5.1.6.1 Plants

Apart from organisms analysed as foodstuffs, only few data exist for beryllium levels in terrestrial or aquatic organisms.

The beryllium contents of plant samples are generally below 1 mg/kg dry weight. Exceptions are some plants that concentrate beryllium from soils (Griffitts et al., 1977). As reported in section 4.2, hickory trees contain as much as 1 mg beryllium/kg dry weight or 30 mg/kg in plant ash. Nikonova (1967) analysed 45 plant species in the Southern Urals (USSR) and detected beryllium in samples of wood, bark, twigs, and leaves of 23 species in concentrations of up to 0.001% (10 mg/kg). While many samples contained only traces of beryllium, samples of birch (*Betula verrucosa* Ehrh.) and larch trees (*Larix sukaczewii* Dylis.) showed elevated concentrations. The use of bioaccumulators as indicators of exploitable ore deposits was discussed by Nikonova (1967).

Conifers usually contain less than 0.1 mg/kg dry weight (< 1 mg/kg in the ash), while dogwood (*Cornus* spp.) and other broad-leaved trees and shrubs contain more than conifers. However, Griffitts et al. (1977) did not give any values.

Investigating the movement of elements into the atmosphere from coniferous trees in the subalpine forests of Colorado and Idaho (USA), Curtin et al. (1974) found beryllium concentrations ranging from traces up to 1 mg/kg in the ash of needles, twigs, and exudate residues. This is well below 0.1 mg/kg on a dry weight basis. The corresponding soil contents were on average 1.4–2.7 mg/kg ash weight (mulch layer), 2.1–5.9 mg/kg (topmost mineral layer), and 1.9–5.4 mg/kg (subsoil).

Dittmann et al. (1984) used leaves of poplar trees (*Populus nigra*) for monitoring beryllium in 2 industrialized regions of the Federal Republic of Germany. In unwashed leaves collected from 38 trees in Saarland in 1982, levels of between 0.002 and 0.2 mg Be/kg dry weight were found compared with 0.2–5 mg/kg dry weight in samples of upper soil layers. The poplar leaves collected in the Ruhr area in 1979 contained beryllium levels of between 0.003 and 0.03 mg/kg dry weight. In samples of grass (*Lolium multiflorum*), from the same environment as the poplar trees, concentrations of between 0.002 and 0.03 mg/kg were found. The considerable local variations in the beryllium contents observed could not be related to any environmental factors.

The same working group analysed spruce needles in the Saarland (Federal Republic of Germany) in 1982 and 1984 (Mueller et al., 1986). One-year-old needles contained between 0.002 and 0.033 mg/kg (mean: 0.013 mg/kg) and 2-year-old needles contained between 0.004 and 0.065 mg/kg dry weight (mean: 0.022 mg/kg).

5.1.6.2 Animals

Meehan & Smythe (1967) analysed various marine organisms and found beryllium levels ranging from non-detectable amounts in eels to 106 µg/kg fresh weight in Cunjevoi tunicates (*Pyura stolonifera*). Byrne & DeLeon (1986) collected samples of oysters (*Crassostrea virginica*) and clams (*Rangia cuneata*) from Lake Pontchartrain, an estuary located in the deltaic plain of the Mississippi River. Oysters contained an average of 51 µg Be/kg dry weight (5.1 µg/kg fresh weight), clams from one site contained 83 µg/kg dry weight (12 µg/kg fresh weight), and those from another site contained 380 µg/kg dry weight (54 µg/kg fresh weight).

Beryllium concentrations in the blubber of bowhead whales (*Balaena mysticetus*) from the western Arctic were below the limit of detection (10 µg/kg) in 6 samples and 10 µg/kg fresh weight in one sample. Of the various tissues analysed, only the liver contained detectable amounts of beryllium(244 µg/kg fresh weight) (Byrne et al., 1985).

Investigating the suitability of bird feathers as bioaccumulation indicators of heavy metals, Mueller et al. (1984) measured between < 5 µg and 50 µg Be/kg dry weight in different feathers of 3 jays (*Garrulus glandarius*), caught in Saarland, Federal Republic of Germany.

5.2 General population exposure

For the general population not exposed to extraordinary sources of beryllium, the principal sources of exposure seem to be food and drinking-water; inhaled air and ingested dust are of minor importance.

The daily human intake of beryllium from food has not been determined, since the data on beryllium levels in food (section 5.1.4) are insufficient for a reliable estimation. In a study in the United Kingdom (Hamilton & Minsky, 1973), the average dietary intake was estimated to be < 15 µg/day. US EPA (1987) calculated a total daily consumption of 422.8 ng, most of which came from food (120 ng/day) and drinking-water (300 ng/day), while air (1.6 ng/day) and dust (1.2 ng/day) reportedly contributed very little to the total intake of beryllium.

Although, from a toxicological point of view, the pulmonary and dermal routes are the decisive routes, the dietary intake of beryllium could explain the "normal" body burden resulting in a urine level of beryllium of approximately 1 μg/litre (section 6.3).

Tobacco smoking is probably a major source of exposure in the general population. Up to 0.074 μg Be/cigarette has been reported to be in the smoke (section 5.1.5), hence, assuming that the smoke is entirely inhaled, an average smoker (20 cigarettes per day) takes in approximately 1.5 μg Be/day.

In the vicinity of point sources (section 5.1.1), the beryllium intake through air and dust can be increased by 2–3 orders of magnitude. In addition, possible secondary occupational or "paraoccupational" exposure of workers' families may significantly increase the beryllium intake through dust, when the clothes of occupationally exposed persons are not kept at the work-place, as was usually the case in the 1940s. Investigating several cases of non-occupational beryllium disease, Eisenbud et al. (1949) conducted air analyses during simulated home laundering of work clothes worn by employees at a beryllium-producing plant. When the clothes were shaken, a short-term beryllium level of 125–2000 μg/m^3 (mean: 500 μg/m^3) was measured in the indoor air. During the whole laundering procedure, an estimated amount of approximately 17 μg/day could be inhaled by a person. Such "neighbourhood cases", i.e., patients in which beryllium disease occurred as a result of indirect exposure outside a plant (NIOSH, 1972), were first reported by Gelman (1938).

Of the various applications of beryllium, only two possible sources of exposure for the general population could be of significance, i.e., mantle-type camping lanterns and some dental alloys (section 3.2). The mantle of a gas lantern contains about 650 μg beryllium in the form of its oxide. The experiments of Griggs (1973) revealed that most of the beryllium (about 400 μg) is volatilized in the first 15 min of burning a new, unused mantle. It was estimated that, in a 14 m^3-camper vehicle, a relatively high short-term concentration of around 18 μg Be/m^3 could occur in the first few minutes. Since the beryllium emission is principally confined to the process of lighting a new mantle, long-term exposure to significant amounts would not be expected.

Because of a potentiated leakage effect, the dissolution of beryllium from dental alloys that contain nickel and beryllium was several orders of magnitude greater than expected (Covington et al., 1985b). After incubation of pieces of dental alloys (squares 1 cm^2 × 0.1 cm) in 5 ml human saliva for 120 days at 37 °C, the saliva contained, depending on the alloy tested, between 0.3 and 3.48 mg Be/litre at pH 6 and between 12.4 and 43.0 mg/litre at pH 2. It should be noted that the normal pH of saliva is between 5.8 and 7.1.

Historically, the use of beryllia in fluorescent tube phosphors, and particularly the disposal of broken tubes, were important sources of exposure to beryllium and resulted in many cases of beryllium disease. The production of beryllium-containing fluorescent tubes was discontinued in 1949, but it appears that they were still in use in Europe after that date (Tepper, 1972).

Emission and exposure standards for beryllium in ambient air in the neighbourhood of beryllium production and processing plants have been established in the USA and USSR. In the USA, beryllium-related industries are required to limit atmospheric emissions of beryllium to 10 g over a 24-h period or to an ambient air level of 0.01 µg/m^3, as a monthly average (US EPA, 1978a). The maximum allowable concentration (MAC) for ambient air, established in the USSR, is 0.01 µg/m^3 (Izmerov, 1985).

USA regulations also require that emissions of beryllium from rocket-motor firing must not exceed 75 µg/(min·m^3) for low-fired (500 °C) beryllium oxide and 1500 µg/(min·m^3) for high-fired (1500 °C) beryllium oxide, both measured within the time limit of 10–60 min, accumulated during any 2 consecutive weeks, at the property line or nearest place of human habitation (Reeves, 1986).

5.3 Occupational exposure

5.3.1 Exposure levels

The range of industrial processes with potential occupational exposure to beryllium has expanded during recent years. Such industries include mining and processing activities, extraction plants, the manufacture of beryllium alloys, beryllium chemicals, and beryllium ceramics, non-ferrous foundries, the manufacture of electronic and aerospace equipment, tools, and dies, metallurgical operations, golf club manufacturing, and other processing of beryllium-containing metals and ceramics. In these industries, beryllium is released during various processes, such as melting, casting, moulding, grinding, buffing, welding, cutting, electroplating, milling, drilling, and baking. In addition, occupational exposure to the dust or fumes of beryllium may occur at rocket-motor test sites, incinerators, and open-burning sites (IARC, 1980; Newland, 1982; Kjellström & Kennedy, 1984; OSHA, Personal communication 1989). In the USSR, beryllium may be released during the production of emeralds (Sidorenko, Personal communication, 1988).

In 1971, approximately 8000 plants were operating in the USA with about 30 000 employees potentially exposed to beryllium, 2500 of whom were involved in the production industry (NIOSH, 1972). Health risks from beryllium exposure can still occur, particularly in the production industry, in which extraction and sintering processes are not easy to control (Preuss & Oster, 1980).

Before 1950, many cases of beryllium disease were caused by the high exposure of workers to beryllium metal and its compounds, during processing and manufacturing activities. Although there is a lack of quantitative data on exposure to beryllium prior to 1947, there seems to be little doubt that extremely high concentrations were encountered in the work-place (NIOSH, 1972).

In the USA, concentrations greater than 1 mg beryllium/m^3 were not unusual (Eisenbud & Lisson, 1983). Laskin et al. (1950) reported dust concentrations of 110–533 μg Be/m^3 during the coke removal operation and 1473–4710 $\mu g/m^3$ during the beryllium-pouring phase. Exposure levels of up to 43 300 $\mu g/m^3$ were reported by Zielinski (1961) for the breathing zone in an alloy plant.

High exposure to beryllium occurred during the production of fluorescent and neon lamps when beryllium was used, together with other metal salts, to coat the inner surface (Kjellström & Kennedy, 1984). After the discovery of an epidemic of beryllium disease among fluorescent lamp workers, this use of beryllium was discontinued in the USA in 1949.

In the USSR, similar exposure conditions were prevalent in the beryllium industry in its early years (Izmerov, 1985). For instance, during the electrolysis of beryllium salts, atmospheric beryllium concentrations between 5 and 700 $\mu g/m^3$ were reported.

In 1949, the US Atomic Energy Commission, one of the largest consumers of beryllium at that time, introduced permissible exposure concentrations of beryllium that were subsequently adopted by the American Conference of Governmental Industrial Hygienists (ACGIH) in 1955 (NIOSH, 1972).

In the 1950s, control measures were installed in beryllium plants to meet occupational standards. As a result, exposure to beryllium was considerably reduced. For example, in an Ohio extraction plant where control measures were applied, the exposure levels were 2 μg Be/m^3 or less in 90-95% of about 2600 air samples analysed (Breslin & Harris, 1959).

However, the US permissible exposure limit of 2 $\mu g/m^3$ was still being exceeded in various plants. In the Ohio plant, 5-10% of the samples still contained concentrations greater than 25 $\mu g/m^3$. In the early 1970s, peak concentrations in a beryllium extraction and processing plant were over 50 times the accepted peak limit value (25 $\mu g/m^3$), i.e., up to 1310 $\mu g/m^3$ (Kanarek et al., 1973). Follow-up analyses in 1974 showed a significant decrease in the exposure levels (Sprince et al., 1978).

In a copper-beryllium alloy production plant, the 2 $\mu g/m^3$ limit was considerably exceeded during the monitoring period between 1953 and 1960, with time-weighted average values of 4.4-9.5 $\mu g/m^3$ in 1953, 9.2-19.1 in 1956, and 23.1-54.6 $\mu g/m^3$ in 1960 (NIOSH, 1972).

A wide range of worker exposure levels is frequently encountered. In a beryllium-alloy plant, concentrations ranged from < 0.1 $\mu g/m^3$ in the mixing areas to 1050 $\mu g/m^3$ in the oxide areas (Cholak et al., 1967). The 5-day average level in this plant was 60.3 $\mu g/m^3$.

The US Atomic Energy Commission presented exposure data from 5 major beryllium-processing plants, for various periods during 1950-61 (NIOSH, 1972). Up to 40-75% of the average breathing-zone concentrations exceeded 2 $\mu g/m^3$, depending on the effectiveness of the control measures installed. Maximum recorded levels were as high as 550 $\mu g/m^3$. However, values exceeding 50 $\mu g/m^3$ were apparently due to the failure of control devices.

The National Institute of Occupational Safety and Health (NIOSH) conducted several air surveys in different beryllium facilities in the USA. Personal air samples taken at factories where machining of beryllium metal and alloys involved drilling, boring, cutting, and sanding operations, did not reveal any detectable amounts of beryllium (Gilles, 1976; Boiano, 1980; Lewis, 1980). In a boat factory where workers were engaged in grid blasting operations, beryllium concentrations of between 6 and 134 $\mu g/m^3$ were measured. This indicates potential overexposure, since respirators were not worn consistently (Moseley & Donohue, 1983). In a beryllium production plant, concentrations of between 0.3 and 160 $\mu g/m^3$ were found in a 1971 survey, the high values appearing in the powdering operations (Donaldson, 1971). In another beryllium production plant, the concentrations of airborne beryllium, as measured in 1972 surveys, rarely exceeded the TLV of 2 $\mu g/m^3$ (Donaldson & Shuler, 1972). Beryllium concentrations in 50 personal samples collected at a secondary copper smelter in 1982 ranged between < 0.2 and 0.5 $\mu g/m^3$ (Cherniack & Kominsky, 1984). In 1983, 121 personal air samples were obtained in the refinery and manufacturing melt areas of a precious metals refinery. The beryllium concentrations ranged from 0.22 to 42 $\mu g/m^3$ (mean 1.4 $\mu g/m^3$) (McManus et al., 1986). Concentrations in the beryllium shop of another plant ranged from < 0.2 to 7.2 $\mu g/m^3$, with a mean of 0.4 $\mu g/m^3$ (Gunter & Thoburn, 1986).

Casting aluminium alloys with 4% beryllium was associated with beryllium air levels in the range of 0.006-0.14 $\mu g/m^3$. The most extensive emission of beryllium occurred during the refining of the alloy. Changes in technology, the exclusion of the refining process, and the installation of ventilation reduced the beryllium concentrations to 0.001–0.004 $\mu g/m^3$ (Naumova, 1967).

A 1983 report showed that compliance with permissible limits was not being consistently attained during the grinding, polishing, cutting, and welding of beryllium-containing alloys in a metal-processing plant in the Federal Republic of Germany (Minkwitz et al., 1983). Breathing-zone air sampling revealed concentrations of between 0.1 and 11.7 μg Be/m^3 in the total dust (sampling duration: 32–133.5 min; analytical method according to NIOSH, 1977). The technical guidance concentrations (TRK) established in the Federal Republic of Germany were exceeded, particularly during the cutting of alloys using a hand cutter (0.1–10.0 μg/m^3) or an automatic cutting machine (1.4–11.7 μg/m^3) and during the welding process (2.1–3.63 μg/m^3 without exhaust extraction; and 1.12–1.34 μg/m^3 with exhaust extraction).

Bobrischev-Pushkin et al. (1975) reported that air contamination, when welding beryllium and its alloys, is determined by the process of welding and depends on the technology and the concentration of beryllium in the materials being used. The highest emissions of beryllium into the air occur during argon-arc welding. During diffusion and electric-beam welding, air contamination occurs while taking the welded objects out of the welding chambers. It is at its highest during the cleaning of the welded objects. The concentration of beryllium in the air can differ greatly, according to its ratio with other metals in the welded alloys. Flux-core welding promotes the liberation of water-soluble beryllium salts and requires protective measures to prevent damage to mucous membranes and the skin.

The exposure of dental laboratory technicians to beryllium, during the processing of beryllium-containing dental alloys, has been investigated. Analyses of air samples collected in the breathing zone showed that 2.0 μg/m^3 was not exceeded when exhaust extraction was used (Dvivedi & Shen, 1983). However, the use of an electric handpiece without exhaust extraction produced a beryllium concentration of 74.3 μg/m^3, compared with a concentration of 1.75 μg/m^3 when exhaust extraction was used.

In the USA, a newly identified use for beryllium-copper alloy has been reported. Workers grinding, polishing, and finishing golf club heads, made from this alloy, were exposed to average beryllium breathing zone concentrations ranging from 2 to 14 μg/m^3 (OSHA, Personal communication, 1989).

To meet industrial hygiene standards, the installation, maintenance and, if necessary, improvement of control measures is required. Engineering controls should provide safe work-place atmospheres, and should include closed systems and special local exhaust devices (Preuss, 1975). Personal protective devices, such as respirators, should be used in cases where high atmospheric concentrations result during emergencies or maintenance or repair (NIOSH, 1972).

5.3.2 Occupational exposure standards

The first hygiene standards for beryllium were introduced by the US Atomic Energy Commission in 1949, on the basis of the recommendations of Eisenbud et al. (1948, 1949). The short-term exposure limit of 25 $\mu g/m^3$, suggested by these authors, was derived from limited investigations (section 9). There was no real empirical basis for the establishment of the long-term occupational exposure level of 2 $\mu g/m^3$, which was derived, not from animal and human data, but by analogy with industrial air limits for the toxicity of other heavy metals (NIOSH, 1972). The values of 2 $\mu g/m^3$, as an 8-h time-weighted average (TWA), 5 $\mu g/m^3$, as a 30-min ceiling limit, and 25 $\mu g/m^3$, as an instantaneous maximum peak exposure limit, never to be exceeded, were adopted in 1971 as the permissible exposure limits in the USA. In addition, beryllium and its compounds are grouped within the group of A2 carcinogens, i.e., industrial substances suspected to have carcinogenic potential for man (ACGIH, 1988).

In the USSR, the occupational exposure limit is 1 $\mu g/m^3$. In addition, maximum allowable levels (MAL) of 2 mg/m^2 for smooth and low-sorbing materials, 5 mg/m^2 for readily sorbing materials, and 0.5 mg/m^2 for the floors of offices and public rooms, have been established (Izmerov, 1985).

Several countries have adopted the exposure limits of either the USA or the USSR (WHO, 1990). In the Federal Republic of Germany, the formerly legally binding MAK value (maximum concentration value in the work-place) of 2 μg Be/m^3 was redefined as a technical guiding concentration (TRK) in 1982, because of the carcinogenic potential of beryllium (DFG, 1988). TRK values are assigned for carcinogenic substances to reduce the risk of a health hazard. A TRK value of 5 $\mu g/m^3$ (calculated as Be in total dust) has been established for grinding beryllium metal and alloys, apparently to meet the technical performance of the associated control measures. For all other beryllium-related work processes, the TRK value is 2 $\mu g/m^3$.

A downward revision of the US standard from 2 $\mu g/m^3$ to 1 $\mu g/m^3$ (TLV) with a 15-min ceiling limit of 5 $\mu g/m^3$, as well as a dermal exposure limit, were proposed by the Occupational Safety and

Health Administration (OSHA, 1975). The proposal also included medical surveillance and other auxiliary provisions, including periodic exposure monitoring, worker training and education, and labelling requirements. As of 1989, the proposed standard had not been implemented.

The National Institute for Occupational Safety and Health (NIOSH) recommended "that occupational exposure to beryllium be controlled so that no workers will be exposed in excess of 0.5 µg Be/m^3" (NIOSH, 1977). As NIOSH classifies beryllium as a carcinogen, control of beryllium exposure to maintain the lowest feasible limit is recommended.

On the basis of animal carcinogenicity studies, it has been suggested in the USSR that the exposure limit in the air of working areas should be 0.01 µg/m^3 (Parfenov, 1988).

5.3.3 Biological monitoring

Analysis of tissues and body fluids for beryllium can indicate previous exposure to beryllium. However, because earlier analytical methods had relatively low sensitivities and were mostly not validated by proper quality control procedures, there is a lack of reliable data that could be used to establish the body burden of beryllium in occupationally exposed people, compared with that in the general population. The "normal" level in urine is elevated following specific exposure, but this is not consistent (section 6.3).

Attempts have been made to use the beryllium contents of the lung or lymph-node specimens as indicators of exposure and body burden. Beryllium storage in tissues is of long duration, especially in pulmonary lymph nodes and bones (section 6.4). As with urinary beryllium, it is not possible, using the limited data available, to establish a clear relationship between exposure and body burden, though clearly elevated levels are found in tissue samples of patients with beryllium disease (section 6.2).

Beryllium levels in lung tissue and urine may be of relevance from the medical point of view, because the detection of elevated beryllium levels in patients with lung disease is indicative of a beryllium-related disease. On the other hand, low levels are not evidence for the absence of chronic beryllium disease. Long-term, low-level exposure and the concomitant elimination of beryllium may result in a relatively low body burden. Moreover, because of the allergic features of some of the effects of beryllium (section 9), it is possible that very low exposures can possibly cause signs and symptoms in previously sensitized persons.

6. KINETICS AND METABOLISM

6.1 Absorption

6.1.1 Respiratory absorption

Inhalation is the primary route of uptake of occupationally exposed persons. There are no data on the deposition or absorption of inhaled beryllium in human beings, but it can be expected that, as with other inhaled particles, dose, size, and solubility are the important factors governing deposition and lung clearance.

Animal studies have shown that, after deposition in the lungs, beryllium is retained and slowly absorbed into the blood. Pulmonary clearance of inhaled beryllium is biphasic, with a fast elimination phase during the first 1–2 weeks after cessation of exposure and a slow elimination phase thereafter. The initial fate of beryllium, deposited in the lung by inhalation or by intratracheal injection, depends on the physical and chemical states of the compounds present.

Van Cleave & Kaylor (1955) studied the kinetics of ^7Be in the rat after intratracheal injection of trace amounts of either ^7Be citrate or $^7BeSO_4$. Absorption of the soluble ^7Be citrate complex from the alveoli into the blood was fast, as can be seen from the observation that almost 79% of the total injected dose had been eliminated in the urine and faeces by day 4 and only 2.5% remained in the lungs, decreasing to 1% at the end of 16 days. The liver and skeleton were the sites of deposition of the mobilized ^7Be. On average, 55% of the original dose of the ^7Be citrate was eliminated via the kidneys in the first 24 h, but only 15% of the $^7BeSO_4$ dose was eliminated in the urine in that time, indicating a much slower absorption from the lungs into the blood. At 16 days, 62% of the total dose had been eliminated in the urine and faeces; 70% of the remaining dose was still present in the lung. Thereafter, pulmonary retention of beryllium decreased with a concomitant increased deposition in mediastinal lymphatic structures. However, in some preparations of $^7BeSO_4$, retention of appreciable amounts was still found after 315 days. This finding probably resulted from the different particulate or colloidal features of the ^{7}Be administered.

Reeves et al. (1967) and Reeves & Vorwald (1967) also observed long retention of beryllium in the lungs of rats following exposure to a BeSO$_4$ aerosol, at a mean concentration of 34 µg Be/m^3, for 72 weeks (7 h/day, 5 days/week). An equilibrium concentration was reached in the lungs and tracheobronchial lymph nodes at about 36 weeks. After cessation of exposure, pulmonary beryllium was first eliminated with a half-time of about 2 weeks, followed by a logarithmically decreasing clearance rate. However, some beryllium (0.2–0.7 µg/kg), remained in the lungs for years, probably in an encapsulated form.

In rats and guinea-pigs, Zorn et al. (1977) observed rapid lung clearance during the first 5 days following a 3-h nasal exposure to BeSO$_4$ tagged with ^{7}BeCl$_2$. Only 2% of the dose of approximately 1634 µg Be per animal remained in the lungs after 5 days. During the following 12 weeks, the pulmonary beryllium content further decreased to 1.5% of the whole body activity, confirming the very slow second elimination phase observed by Reeves & Vorwald (1967).

The relevance of the test material with respect to the clearance of beryllium from the lungs of rats can also be demonstrated by the results of several other studies. Kuznetzsov et al. (1974) found a pulmonary half-time of 20 days following intratracheal administration of ^{7}BeCl$_2$ (10 µCi). Bugryshev et al. (1976) found that 20% of an intratracheal dose of BeCl$_2$ was retained in the lungs of rats, after 94 days. Significant pulmonary retention of low-solubility BeO (unknown specification), which lasted for several months, with a retention half-life of approximately 12 months, was found in rats by Dutra et al. (1951). Sanders et al. (1975) and Rhoads & Sanders (1985) found alveolar half-times in rodents of about 6 and 13 months, respectively, after inhalation of high-fired (1000 °C) BeO.

Most beryllium that is inhaled is in the form of particulate matter and must be mobilized by other means than passing directly into the blood. This is mainly accomplished through the mucociliary transport of particles deposited in the tracheobronchial tree, (which, according to Camner et al. (1977), is probably responsible for the rapid clearance of inhaled particles during the first day), and also through the uptake of beryllium by alveolar macrophages. Hart & Pittman (1980) observed that insoluble beryllium complexes were more effectively incorporated *in vitro* than soluble compounds. The temperature and energy dependency that was observed, together with the demand for calcium and magnesium ions, suggests that phagocytosis was involved.

The results of studies by Lundborg et al. (1984) and Nilsen et al. (1988) showed that alveolar macrophages dissolved inorganic particles, probably because of the low pH in the phagolysosomes. André et al. (1987) found that the dissolution by alveolar macrophages of two industrial forms of beryllium, i.e., particles of metal powder and particles of hot-pressed beryllium, was consistent with the *in vivo* clearance observed in rats and baboons.

Beryllium-coated particles were highly toxic to macrophages (Camner et al., 1974), indicating that elevated beryllium concentrations inhibit pulmonary clearance. The decrease in the alveolar clearance of a test aerosol (^{239}PuO$_2$) to 60% of the normal rate, after exposure of rats to an aerosol of BeO was probably due to the inhibition of macrophages by incorporated beryllium (Sanders et al., 1975).

Lung clearance of beryllium appears to be species and sex-dependent. Clearance was more rapid in hamsters than in rats, and, in both species, it was greater in males than in females (Sanders et al., 1975). Reeves & Vorwald (1967) also observed a greater clearance rate in male rats than in females.

6.1.2 Dermal absorption

Uptake of beryllium through skin absorption contributes only very small amounts to the total body burden of beryllium-exposed persons. However, because of the skin effects elicited by beryllium compounds, this route is of some significance.

Trace levels of ^{7}BeCl$_2$ were found to be absorbed at a low rate through the rat tail (Petzow & Zorn, 1974). Although subsequent systemic distribution of trace amounts of beryllium was observed, systemic distribution through the intact skin is not expected following local contact, because most beryllium salts do not remain soluble at physiological pH (Reeves, 1986). Belman (1969) studied the beryllium-binding properties of epidermal constituents of guinea-pigs and found that ionic beryllium applied to the skin became bound mainly to alkaline phosphatase and nucleic acids.

6.1.3 Gastrointestinal absorption

Faecal elimination of beryllium following its uptake through inhalation (section 6.1.1) indicates that part of the inhaled material goes to the gastrointestinal tract, either by mucociliary action or by the swallowing of the insoluble material deposited in the upper respiratory tract (Kjellström & Kennedy, 1984).

Estimates from animal studies, in which trace amounts of carrier-free ^{7}Be were administered as the chloride, show that absorption of ingested beryllium is very low, with values generally below 1% in pigs (Hyslop et al., 1943), rats, mice, monkeys, dogs (Crowley et al., 1949; Furchner et al., 1973) and cows (Mullen et al., 1972).

From the data reported by Reeves (1965), it appears that, in rats, the absorption rate for $BeSO_4$ is much higher, since about 80% of the ingested beryllium (doses: 0.6 and 6.6 µg Be/day in the drinking-water) was eliminated with the faeces. It was assumed that the remainder was absorbed from the stomach, at the pH of which the $BeSO_4$ is in the ionized form; at the alkaline pH of the intestine, beryllium precipitates as the phosphate. However, in contrast to other studies, Reeves (1965) did not use radionuclide ^{7}Be as tracer, but analysed beryllium by a spectrometric method. The recovery was only between 60 and 91% of the total consumption. On subtracting the measured beryllium content of the gastrointestinal tract from the total body burden plus the urinary beryllium, it is evident that only 0.06–1.5% of the total intake must have been absorbed from the gastrointestinal tract into the blood and distributed to the tissues or excreted by the kidneys.

6.2 Distribution and retention

Most of the beryllium circulating in the blood is transported as colloidal phosphate, and only small amounts are transported as citrate or hydroxide (Reeves, 1986). Stiefel et al. (1980b) found that beryllium is bound to the prealbumins and γ-globulin. The relative distribution in these two fractions depends on the concentration of beryllium in the blood. At 1 µg/kg, 8% is stored in the prealbumin and about 60% in the γ-globulin. Above 10 µg/kg, the distribution is reversed.

There is also a binding site for beryllium on the lymphocyte membrane (Skilleter & Price, 1984).

Regardless of the animal species, a significant part of the beryllium administered is incorporated in the skeleton. The extent to which beryllium is deposited in other target tissues depends on the route of administration, and the physical and chemical state, and dose of the compound.

Following oral administration of carrier-free $^7BeCl_2$ (0.7–1.3 mCi) to calves, only about 0.3% of the dose was absorbed from the gastrointestinal tract, of which approximately 90% wase found in the bone. The remainder was mainly localized in the gastrointestinal tract and only 2% of the total body burden was found in the liver (Mullen et al., 1972).

In rats receiving an average daily dose of 6.6 µg Be per rat ($BeSO_4$ in the drinking-water) for 24 weeks, 76% of the total body recovery, minus the beryllium content in the gastrointestinal tract, was in the skeleton, 16% in the blood, and 7% in the liver. At a dose of 66 µg/day, 85% of the beryllium retained was deposited in the skeleton, smaller amounts being found in the blood (11%) and the liver (3%) (Reeves, 1965). The doses administered in this study were a factor of 1000 higher than those in the preceding study.

Scott et al. (1950) observed that carrier-free ^7Be, injected intravenously in rats and rabbits, was mainly deposited in the bone.

Less than 10% of an intravenous dose of $BeSO_4$ (0.75–15 µg ^7Be/kg body weight) was found in the liver of rats, 24 h after administration, but more than 25% was found at doses of 63 µg/kg body weight or more (Witschi & Aldridge, 1968).

The distribution of intratracheally injected BeO is highly dependent on the properties of the compound used (Spencer et al., 1965). Relatively high levels of beryllium were found in the liver, kidney, and bone of rats that had been treated with low-fired BeO (produced by calcining α-$Be(OH)_2$ for 10 h at 500 °C). In contrast, beryllium from particles of high-fired BeO (1600 °C), which has a low solubility, remained mainly in the lungs and very little was distributed to the liver, kidneys, and bone. However, no values were given by Spencer et al. (1965). Rhoads & Sanders (1985) did not detect beryllium at any extrapulmonary site in rats following nose-only inhalation of BeO dust (1000 °C). Clearance to the pulmonary lymph nodes was about 2%, 63 days after exposure.

In general, inhalation exposure to beryllium compounds results in long-term storage of appreciable amounts of beryllium in lung tissue, particularly in pulmonary lymph nodes (section 6.1.1), and in the skeleton, which is the ultimate site of beryllium storage. More soluble beryllium compounds are also translocated to the liver, abdominal lymph nodes, spleen, heart, muscle, skin, and kidney.

Bencko et al. (1979a) reported that the placental permeability for soluble $^7BeCl_2$ (0.1 mg/kg body weight), intravenously administered to mice, was slight. The concentrations of beryllium in the placenta and in the remaining organs of the females was one order of magnitude higher than those in the fetuses. The transfer of ingested radioactive beryllium (3.1 mCi of carrier-free $^7BeCl_2$ per animal) to the milk was low (Mullen et al., 1972). Less than 0.002% of the administered activity was secreted in the milk of cows.

Apart from bone, the lung is considered to be the primary target organ in man. Sprince et al. (1976) analysed specimens taken at autopsy and found less than 20 μg Be/kg dry weight in lung tissue (mean: 5 μg/kg; range: 3–10 μg/kg; 6 cases) and mediastinal lymph nodes (mean: 11 μg/kg; range: 6–19 μg/kg; 7 cases) of control patients without granulomatous disease. These levels are in agreement with the normal range of 2–30 μg Be/kg dry lung tissue covering 90% of the values found in 125 lung specimens obtained during thoracic surgery (Baumgardt et al., 1986).

Using inductively coupled plasma atomic emission spectrometry, Caroli et al. (1988) analysed different parts of lung tissue of 12 subjects in the urban Rome area. All were non-smokers, 50 or more years old, and had not been occupationally exposed to beryllium during their life-time. The overall mean of 5 μg/kg fresh weight (median 6 μg/kg; 9th percentile 8 μg/kg) indicates a far smaller concentration range than those above, which were on a dry weight basis.

Analysis of lung tissue from 66 patients with beryllium disease showed that 82% had Be levels of more than 20 μg/kg dry weight. Even higher levels were found in lymph-node specimens from 5 patients; the peripheral lymph nodes contained between 2 and 490 μg Be/kg dry weight (mean: 110 μg/kg), the mediastinal nodes contained between 56 and 8500 μg/kg (mean: 3410 μg/kg) (Sprince et al., 1976).

From the examination of cases from the US Beryllium Case Registry (Freiman & Hardy, 1970), it appears that the levels of beryllium in the lungs of patients dying with acute disease are generally higher than those in patients with chronic disease. As expected, neither lung nor urinary beryllium levels are correlated with the occurrence or severity of chronic beryllium disease (Tepper, 1972). Apart from considerable variations in the beryllium concentrations of several samples from the same lung, great variability also exists in the tissue levels of beryllium in the patients. There are more people with high beryllium body burdens and no beryllium disease than there are people with chronic beryllium disease. For example, healthy refinery workers had 1000 times higher values than persons with beryllium disease (Tepper, 1972).

6.3 Elimination and excretion

As pointed out in section 6.1, elimination of absorbed beryllium occurs mainly in the urine and only to a minor degree in the faeces. Most of the beryllium taken up by the oral route passes through the gastrointestinal tract unabsorbed and is eliminated in the faeces.

Rats injected intramuscularly with carrier-free ^{7}Be (approximately 20 mCi, i.e., 57 ng ^{7}Be per rat) eliminated 15%, 14.6%, 24.4%, and 44% of the dose in the urine at 1, 4, 16, and 64 days, respectively, versus 4.25%, 4.17%, 9.25%, and 13.1% in the faeces (Crowley et al., 1949).

After intravenous administration of very small doses of carrier-free ^{7}Be to rats (0.09 ng Be/kg body weight) and rabbits (0.04 ng Be/kg body weight), urinary excretion was the major elimination route (Scott et al., 1950). Elimination was greatest during the first 24 h and amounted to 38.8% of the total dose in rats and 28.8% in rabbits. In comparison, animals receiving ^{7}Be plus $BeSO_4$ as a carrier, and, thus, a relatively large dose of 0.15 μg Be/kg body weight (rats) or 0.05 μg/kg body weight (rabbits), excreted only 24.2% (rats) or 14% (rabbits) of the dose in the urine. This reduction of the urinary excretion rate with increasing dose may be explained by the increasing immobilization of beryllium, because of its binding on proteins. Faecal elimination of beryllium was comparatively low during the first day, with only 0.1% of the dose excreted by this route in rabbits and 3.5% (carrier-free ^{7}Be) or 4.2% (with carrier) in rats. Following the first rapid urinary elimination phase, the daily urinary elimination in rabbits varied between 0.5 and 1.8% of the dose with a concomitant faecal elimination of 0.2–0.5%.

This observation is consistent with the results of Furchner et al. (1973) who determined urinary/faecal ratios of 3.21 in mice and 10.2 in rats, during the first 24 h after intraperitoneal administration, and 3.5 in mice, 21.34 in rats, 4.03 in monkeys, and 48.61 in dogs after intravenous administration. Thereafter, the high urinary excretion rate declined rapidly and the amount lost in the faeces equalled that in the urine. The mechanism of urinary excretion is probably active tubular secretion, because most of the colloidally bound plasma beryllium does not pass the glomerulus in the kidney (Reeves, 1986).

True biliary excretion seems to play a minor role in total beryllium elimination (Cikrt & Bencko, 1975). Elevated amounts of beryllium eliminated in the faeces after intratracheal or inhalation administration are probably the result of clearance from the respiratory tract and ingestion of swallowed beryllium.

Quantitative data on the excretion of beryllium in human beings are confined to some urinary levels in exposed and non-exposed people.

Twenty of 22 non-occupationally exposed persons living in the vicinity of beryllium plants did not have any beryllium (< 0.02 μg/litre) in their urine, though 8 of them were suspected of, or diagnosed as, having berylliosis. Two persons living near a beryllium refinery (0.4 km distance) showed urine concentrations of 0.02 and 0.06 μg Be/litre, respectively (Lieben et al., 1966). Grewal & Kearns (1977) found an average concentration of 0.9 ± 0.4 μg/litre in 120 people from California. A similar value (0.9 ± 0.5 μg/litre) was reported by Stiefel et al. (1980a) for 20 non-occupationally exposed persons from the Federal Republic of Germany. It appears that a "normal" beryllium level in urine is around 1 µg/litre. However, this level seems much too high considering that gastrointestinal absorption of beryllium into the blood is very low. From the estimated inhalation intake of 1.6 ng Be/day per person it can be assumed that only a few nanograms of beryllium will be excreted daily (US EPA, 1987).

The discrepancy between the reported and expected urinary beryllium levels cannot be explained. The contribution of food is unclear, since reliable data are not available (section 5.1.4). Human data on the bioavailability of ingested beryllium are also lacking. Moreover, the studies have not distinguished between smokers and non-smokers. Stiefel et al.(1980a) reported levels of about 2 µg Be/litre in the urine of smokers.

An increase in urinary beryllium of several µg/litre, following inhalation exposure to beryllium, was reported by Hardy & Chamberlin (1972). In the urine of 8 laboratory assistants, the beryllium concentration increased from 1 µg/litre to about 4 µg/litre, when the beryllium concentration in the laboratory air increased from about 0.4 ng/m^3 to 8 ng/m^3, because of accidental contamination with BeCl$_2$ (Stiefel et al., 1980a). However, Lieben et al. (1966) only found levels ranging between non-detectable (< 0.02 µg/litre) and 0.26 µg/litre in the urine of beryllium workers.

6.4 Biological half-life

A distinction must be made between the elimination of inhaled beryllium from the lungs, and the total elimination of beryllium from the body. In the first case, studies indicate that only the non-ionized soluble forms of beryllium, such as the citrate, are cleared from the lung rapidly (in about 4 days). The ionized soluble forms become precipitated in lung tissue and behave like particulate matter. Their clearance consists of a "fast phase" and a "slow phase". The fast phase is probably because of uptake in macrophages, which subsequently migrate out of the bronchopulmonary system (Van Cleave & Kaylor, 1955; Kuznetsov et al., 1974; Hart & Pittman, 1980; Hart et al., 1984; Finch et al., 1986). The half-time of the fast phase is in the range of 1–60 days (Sanders et al., 1975; Rhoads & Sanders, 1985). The slow phase of beryllium clearance has a half-time of 0.6–2.3 years and it may represent the slow dissolution and dissipation of the deposits that have either become encapsulated in scar tissue or otherwise rendered unavailable to the phagocytic action of migratory cells (Reeves, 1968; Rhoads & Sanders, 1985; Finch et al., 1986). There appears to be a sex difference in the efficiency of clearance, at least in rats, favouring males compared with females (Reeves & Vorwald, 1967; Reeves, 1968).

After intravenous injection of carrier-free $^7BeCl_2$, Furchner et al. (1973) calculated biological half-lives of 1210, 890, 1770, and 1270 days in mice, rats, monkeys, and dogs, respectively.

In human beings, the residence time for beryllium in the lung may be several years, since appreciable amounts of beryllium can be found in people, many years after cessation of exposure to beryllium (section 5.3.3). In a report of the International Commission on Radiological Protection (ICRP, 1960), the biological half-life of beryllium in human beings was calculated to be 180, 120, 270, 540, and 450 days in the total body, kidneys, liver, spleen, and bone, respectively.

7. EFFECTS ON ORGANISMS IN THE ENVIRONMENT

7.1 Microorganisms

Little is known about the effects of beryllium on microorganisms. In an earlier study, Pirschle (1935) noted a marked stimulating effect of higher concentrations of $BeCl_2$ (0.1–0.001 mol/litre) and $BeSO_4$ (0.1–0.01 mol/litre) on mycelial growth. $Be(NO_3)_2$ did not affect growth, but suppressed the formation of conidia. The effects of beryllium were more comparable with those of aluminium than with those of the other alkaline earths, reflecting its specific chemical properties (section 2.2).

Gormley & London (1973) performed various experiments using mixed and pure soil microorganisms grown in media containing 100 mg/litre of $BeSO_4.4H_2O$ complexed with sodium citrate. They did not observe any inhibitory effects of beryllium on cell growth. In one mixed culture, a 20-h delay before the onset of the log phase of growth occurred. However, in this culture, a higher yield of biomass was noted compared with the control. Soil microorganisms, grown in a magnesium-deficient medium, grew better in the presence of beryllium, indicating that beryllium can substitute magnesium to some extent. This effect has also been observed in plants (section 7.3) and could be responsible for the stimulatory effects reported.

Studies on *Pseudomonas aeruginosa* showed that several growth factors are affected by potassium dioxalatoberyllate ($K_2[Be(C_2O_4)_2]$) at concentrations of 8 µmol/litre (72 µg/litre) or more. Also the production of the pigment pyocyanin was found to be inhibited (MacCordick et al., 1976).

Bringmann & Kuehn (1981) determined the growth-inhibiting effects of beryllium nitrate, $Be(NO_3)_2.4H_2O$, in protozoa. The toxicity threshold levels were 0.004 mg Be^{2+}/litre for the flagellate *Entosiphon sulcatum* Stein, 0.017 mg/litre for the ciliate *Uronema parduczi* Chatton-Lwoff, and 0.51 mg/litre for the flagellate *Chilomonas paramaecium* Ehrenberg.

Wilke (1987) investigated the effects on soil microorganisms of $BeSO_4$ added to fertilizers. At a concentration of 30 mg Be/kg soil, the biomass was reduced to 60% and nitrogen-mineralization to 57% of the control. At 80 mg/kg, the activities of dehydrogenase, saccharase, and protease were also inhibited, while ATP-content, alkaline phosphatase activity, and nitrification were unaffected.

7.2 Aquatic organisms

7.2.1 Plants

Hoagland (1952a) found that $BeSO_4.4\ H_2O$, at concentrations of 2×10^{-4} to 3×10^{-4} mol Be/litre (1.8–2.7 mg/litre), inhibited growth of the green alga *Chlorella pyrenoidosa* by only 5.6 ±5.9% at an initial pH of 11.4, which decreased to about pH 7 in 24 h. In further experiments, Hoagland (1952a) observed that, at a low initial pH of 6.3, growth of both magnesium-deficient (1×10^{-4} mol Mg/litre) and high-magnesium algae (2×10^{-3} mol Mg/litre) was depressed by the addition of beryllium (2×10^{-4} mol Be/litre). However, at pH 11.4, beryllium had a stimulatory effect, probably because it became available to the algae and substituted magnesium in the growth process, but not in the demands of chlorophyll.

Karlander & Krauss (1972) showed that the growth of *Chlorella vannieli* was inhibited by $BeCl_2$ at a concentration of 100 mg Be/litre.

7.2.2 Animals

Laboratory studies on the acute toxicity of beryllium for freshwater species are summarized in Table 9. Only one invertebrate species (*Daphnia magna*) has been studied. Some 48-h EC_{50} values were 7.9 mg Be/litre for $BeCl_2$ and 18.0 mg Be/litre for $Be(NO_3)_2$. In fish species, LC_{50} values varied from 0.15 to 32.0 mg Be/litre, depending on the species and test conditions.

Beryllium sulfate was one to two orders of magnitude more toxic for fathead minnows and bluegills in soft water than in hard water (Tarzwell & Henderson, 1960). Slonim & Slonim (1973) noted an exponential increase in the toxicity of beryllium for guppies, with decreasing hardness.

Salamander larvae showed a similar sensitivity to beryllium (Table 9) and were also more adversely affected in soft than in hard water (Slonim & Ray, 1975).

The effects of beryllium on the development of early-life stages have been examined in few studies. Dilling & Healey (1926) examined the germination of frog spawn and the growth of tadpoles. Beryllium nitrate at concentrations of 0.9–4.5 mg Be/litre did not interfere with the development of eggs of undefined frog species and tadpoles grew well at concentrations of 0.09–0.2 mg Be/litre. Hildebrand & Cushman (1978) did not observe any adverse effects on the development of eggs of the carp (*Cyprinus carpio*) at beryllium concentrations below 0.08 mg/litre. However, concentrations above 0.2 mg/litre reduced hatching success to 0%. The hardness of the spring water used was approximately 50 mg $CaCO_3$/litre and carp eggs responded only slightly more sensitively than adult fish under these low hardness conditions.

US EPA (1980) cited a comparative toxicity study on *Daphnia magna*. The 48-h EC_{50} and chronic toxicity values in the same test water (hardness: 220 mg $CaCO_3$/litre) were 2500 and 5.3 μg Be/litre, respectively, indicating a large difference between acute and chronic toxicity. No effects on reproduction were observed at 3.8 μg/litre.

7.3 Terrestrial organisms

7.3.1 Plants

In studies with controlled nutrient media, Hoagland (1952a) demonstrated a definite relationship between the presence of the chemically similar magnesium and the effects of beryllium on the growth of tomato plants. At a pH above 9, addition of 2×10^{-4} mol Be/litre (1.8 mg/litre) as $BeSO_4.4\,H_2O$ to magnesium-deficient solutions produced rapid growth without evidence of magnesium deficiency; in the absence of beryllium, growth was depressed and chlorosis occurred within 2 weeks. It seems that beryllium can reduce the magnesium requirement of plants, but not absolutely, as plants with higher levels of magnesium deficiency grew at a slower rate and died with no sign of chlorosis. The pH-dependency of

Table 9. Acute toxicity of beryllium for freshwater animals

Test species	Test type	Test chemical	Hardness (mg/litre as $CaCO_3$)	Test duration (h)	Effect	Concentration (mg Be/litre)	Reference
Water flea (*Daphnia magna*)	static	beryllium nitrate	300	24	LC_{50}	18	Bringmann & Kühn (1977)
Water flea (*Daphnia magna*)	static	beryllium chloride	180	48	EC_{50}	7.9	US EPA (1978b)
Bluegill (*Lepomis macrochirus*)	static	beryllium sulfate	20	96	LC_{50}	1.3	Tarzwell & Henderson (1960)
Bluegill (*Lepomis macrochirus*)	static	beryllium sulfate	400	96	LC_{50}	12	Tarzwell & Henderson (1960)
Brook trout (*Salvelinus fontinalis*)	static	beryllium sulfate	140	96	LC_{50}	5	Cardwell et al. (1976)
Channel catfish (*Ictalurus punctatus*)	static	beryllium sulfate	140	96	LC_{50}	5	Cardwell et al. (1976)
Fathead minnow (*Pimephales promelas*)	flow-through	beryllium sulfate	140	96	LC_{50}	3.25	Cardwell et al. (1976)
Fathead minnow (*Pimephales promelas*)	static	beryllium sulfate	20	96	LC_{50}	0.15-0.2	Tarzwell & Henderson (1960)
Fathead minnow (*Pimephales promelas*)	static	beryllium sulfate	400	96	LC_{50}	11-20	Tarzwell & Henderson (1960)

Table 9 (continued)

Test species	Test type	Test chemical	Hardness (mg/litre as CaCO$_3$)	Test duration (h)	Effect	Concentration (mg Be/litre)	Reference
Flagfish (Jordanella floridae)	flow-through	beryllium sulfate	140	96	LC$_{50}$	3.5-4.4	Cardwell et al. (1976)
Goldfish (Carassius auratus)	flow-through	beryllium sulfate	147	96	LC$_{50}$	4.8	Cardwell et al. (1976)
Guppy (Poecilla reticulata)	static	beryllium sulfate	22	96	LC$_{50}$	0.16	Slonim & Slonim (1973)
Guppy (Poecilla reticulata)	static	beryllium sulfate	150	96	LC$_{50}$	6.1	Slonim & Slonim (1973)
Guppy (Poecilla reticulata)	static	beryllium sulfate	275	96	LC$_{50}$	13.7	Slonim & Slonim (1973)
Guppy (Poecilla reticulata)	static	beryllium sulfate	400	96	LC$_{50}$	20	Slonim & Slonim (1973)
Guppy (Poecilla reticulata)	static	beryllium sulfate	450	96	LC$_{50}$	19-32	Slonim & Slonim (1973)
Salamander larvae (Ambystoma maculatum)	static	beryllium sulfate	22	96	LC$_{50}$	3.2-8.3	Slonim & Ray (1975)
Salamander larvae (Ambystoma maculatum)	static	beryllium sulfate	400	96	LC$_{50}$	18-31	Slonim & Ray (1975)

these phenomena raises the question as to whether the beryllate ion (BeO_2^{--}), formed above a pH of 8, is the biologically active agent.

At more acidic pH values and at higher bioavailable concentrations, beryllium is phytotoxic. Romney et al. (1962) noted a definite decrease in the total dry weights of bush beans (*Phaseolus vulgaris*) grown in nutrient solutions at a controlled pH of 5.3. The mean total dry weights were 60.2, 40.2, 35.5, 20.6, 14.5, and 7.3 g from the 0, 0.5, 1, 2, 3, and 5 mg Be/litre cultures, respectively. Reduction of yield was also seen in soil cultures of beans, wheat, and ladino clover at beryllium levels corresponding to 4% of the cation-exchange capacity in soil (Romney & Childress, 1965). Effects were first observed on the roots, which turned brown and failed to resume normal elongation. It should be noted that roots accumulated most of the beryllium taken up, and very little was translocated to the upper parts of the plants. Stunting of both roots and foliage occurred, but there was no chlorosis or mottling of foliage.

Williams & Le Riche (1968) observed similar effects and reduced yield in kale grown in a nutrient culture solution containing more than 2 mg/litre of beryllium (as $BeSO_4.4 H_2O$). However, at a concentration of 0.5 mg/litre, the yield was greater than in the control.

Hara et al. (1977) grew cabbage plants (*Brassica oleracea* L. var. *capitata* L.) in culture solutions containing 0, 0.5, 5, or 25 mg Be/litre (as $Be(NO_3)_2$) with a low or high supply (20 or 200 mg/litre) of calcium. The dry weights of each part of the plant, especially of the inner leaves, decreased with increasing levels of beryllium. A low calcium content increased this effect. The critical content of beryllium that resulted in a 50% decrease in yield was estimated to be about 3000 mg/kg in the roots and 6 mg Be/kg dry weight in the outer leaves. The latter value corresponded well with the "upper critical level" of 0.6 mg Be/kg dry weight in the leaves and shoots of spring barley, *Hordeum vulgare* L. (Davis et al., 1978).

In soil culture, beryllium phytotoxicity is governed by the nature of the soil, particularly its cation-exchange capacity, and the pH of the soil solution. Romney & Childress (1965) found that beryllium was strongly adsorbed by soils and bentonite, but not by kaolinite. It displaced barium, calcium, magnesium, and strontium in various soil types and in bentonite. With increasing acidity, beryllium became more soluble and hence more toxic to plants. Williams &

Le Riche (1968) concluded that the diminished phytotoxicity under alkaline conditions was the result of precipitation of beryllium as a phosphate salt, making it unavailable to plants.

Kick et al. (1980) studied the effects of beryllium, administered as $BeCl_2$, on the yields of plants. Beryllium at 10 mg/kg sandy soil reduced the yield of spring barley (kernels) by about 26%. Addition to peat led to a yield reduction of 72–79%, whereas addition of kaolin diminished the yield-depressing effect of $BeCl_2$.

Beryllium also suppressed the germination of cress seed (*Lepidium sativum* L.) at concentrations above 10^{-3} mol Be/litre (as $BeCl_2$ and $BeSO_4.4H_2O$) at a pH of 5–6. In addition, pigment analyses showed a reduction in chlorophyll content, which, however, was not correlated with beryllium concentrations (10^{-7}–10^{-3} mol/litre) in the solutions applied (Langhans, 1984).

The mechanism underlying the phytotoxicity of beryllium probably involves its inhibitory effects on enzyme activity and on the uptake of essential mineral ions. As has been shown with animal phosphatases (section 8.7.1), beryllium in micromolar amounts also inhibits plant phosphatases (Hoagland, 1952b). Romney & Childress (1965) noted inhibition of ribulose 1.5-diphosphate carboxylase and phosphoenolpyruvate carboxylase at concentrations above 1 μmol $Be(NO_3)_2$/litre. The resulting interference with phosphorus metabolism is reflected by the enhanced phosphorus uptake observed in pea plants (Lebedena, 1960) and increased phosphorus concentrations in the tissues of alfalfa, barley, pea, and lettuce plants (Romney & Childress, 1965). Conversely, uptake of calcium was reduced in all plant parts, particularly in the roots. Uptake of Na, K, Fe, and Mn was not influenced in these plants. However, in bush beans grown in nutrient solutions, leaf concentrations of these elements and of Cu, Zn, B, Al, Si, Mo, Sr, and Ba were decreased by high beryllium concentrations (8–16 mg/litre as $BeCl_2$) (Romney et al., 1980).

Encina & Becerra (1986) studied the effects of beryllium on cytokinesis in onion root tips. $BeCl_2$ at concentrations ranging from 3–10 mmol/litre was found to slightly inhibit cytokinesis. Induction of binucleate cells attained 2% at 10 mmol/litre. At higher concentrations, production of binucleate cells clearly decreased, probably because beryllium slowed down the rate of telophase and the mitotic index. Beryllium may displace calcium from its binding site, hence prohibiting the formation of cell plates. This hypothesis was confirmed by the finding that, at higher calcium levels, the specific inhibiting effect of beryllium was negated.

7.3.2 Animals

No data are available on the effects of beryllium on domestic or wild terrestrial animals in the environment.

8. EFFECTS ON EXPERIMENTAL ANIMALS AND *IN VITRO* TEST SYSTEMS

8.1 Single exposures

The acute toxicity of beryllium was first described by Siem (1886) who observed that the toxicity of subcutaneously injected beryllium in cats, dogs, and rabbits was ten times higher than that of aluminium.

Acute toxicity data for different beryllium compounds in experimental animals are summarized in Table 10. The LD_{50} data show the low toxicity of ingested beryllium, compared with that of parenterally administered beryllium. This is because of intestinal precipitation of beryllium as nonabsorbable phosphate.

Signs of acute beryllium poisoning, which were observed in LD_{50} studies, were respiratory disorders, spasms, hypoglycaemic shock, and respiratory paralysis (Kimmerle, 1966). Hypoglycaemia was attributable to liver necrosis caused by beryllium. Aldridge et al. (1949) observed severe midzonal liver necrosis, 1–2 days after administration of lethal doses of beryllium.

Melnikov (1959) determined an LC_{50} of 3 mg Be/m^3 for mice, following a 2-h inhalation exposure to beryllium acetate mist. Signs of poisoning were coughing, inflammation of the mucous membranes, dyspnoea, and marked cyanosis before death. Pathological examination revealed diffuse oedema in lung tissue, occasional desquamative pneumonia, and marked degenerative and proliferative changes in the liver, kidneys, and spleen.

The toxicity of ash from the burning of coal (section 3.1) with a relatively high beryllium content was described by Jirele et al. (1966). This initiated a series of studies concerned with the immunotoxicity and toxicokinetics of this metal in Czechoslovakia.

By means of cell kinetic studies and histopathological examination, Sendelbach et al. (1986) determined the acute response in the lungs of mice and rats exposed to an aerosol of BeSO$_4$ (13 mg/m^3, 1 h).

Table 10. Acute toxicity of beryllium compounds (LD_{50} values expressed as mg Be/kg body weight)

Compound/ animal	Route of administration[a]	LD_{50}	Reference
Beryllium acetate			
Rat	ip	22.4	Venugopal & Luckey (1978)
Beryllium carbonate			
Guinea-pig	ip	1.2	Venugopal & Luckey (1978)
Beryllium chloride			
Guinea-pig	ip	6.3	Cochran et al. (1951)
Mouse	im	1.3	Venugopal & Luckey (1978)
Mouse	ip	0.15	Bianvenu et al. (1963)
Rat	ip	0.6	Cochran et al. (1951)
Rat	oral	9.8	Venugopal & Luckey (1978)
Beryllium fluoride			
Mouse	iv	0.34	Kimmerle (1966)
Mouse	oral	19.1	Kimmerle (1966)
Mouse	sc	3.8	Kimmerle (1966)
Beryllium hydroxide			
Rat	iv	0.8	Venugopal & Luckey (1978)
Beryllium nitrate			
Guinea-pig	ip	3.48	Hyslop et al. (1943)
Mouse	iv	0.5	Kimmerle (1966)
Mouse	sc	10.8	Kimmerle (1966)
Beryllium phosphate			
Mouse	iv	1.4	Venugopal & Luckey (1978)
Rat	iv	0.36	Venugopal & Luckey (1978)
Rat	oral	6.5	Venugopal & Luckey (1978)

Table 10 (continued)

Compound/ animal	Route of administration[a]	LD_{50}	Reference
Beryllium sulfate			
Chicken	iv	4.8	Krampitz et al. (1978)
Chicken	sc	3.7	Krampitz et al. (1978)
Monkey	iv	0.05	Venugopal & Luckey (1978)
Mouse	iv	0.04	White et al. (1951)
Mouse	oral	6.95	Venugopal & Luckey (1978)
Mouse	sc	0.13	Morimoto (1959)
Rat	ip	1.54	Sutton (1939)
Rat	iv	0.62	Scott (1948)
Rat	oral	7.02	Venugopal & Luckey (1978)
Rat	sc	0.13	Morimoto (1959)

[a] im = intramuscular; ip = intraperitoneal; iv = intravenous; sc = subcutaneous.

The animals did not exhibit any external effects during the observation period of 21 days. On histopathological examination, rat lungs showed three times greater cellular proliferation than those of mice. The differential cell counts indicated that there was endothelial and epithelial cell injury and an increase in interstitial cells in rats, whereas the predominating cell populations in mice were alveolar macrophages and interstitial and endothelial cells. Histopathological changes were also more severe in rats. The interstitium was thickened with infiltrated interstitial macrophages and segmented leukocytes. Three weeks after exposure, this response was largely resolved.

Bencko et al. (1979b) induced experimental berylliosis in female rats. Five months after intratracheal instillation of 0.1 mg of beryllium oxide, the lesions were confirmed by histopathological study, and by positive migration inhibition tests on macrophages derived from splenic fragments. When the animals were mated, 6 weeks after administration, and the F_1 generation was tested for genetic transmission of beryllium hypersensitivity, the results of the migration inhibition test were negative.

Rats and hamsters were exposed to an aerosol of beryllium oxide (BeO) calcined at 1000 °C (Sanders et al., 1975). The duration of single nose-only exposures ranged from 30 to 180 min. Eight months after exposure, some granulomatous lesions were seen in rats exposed to the highest dose (100 µg Be/m^3). The macrophages assumed a foamy appearance within a month of exposure. BeO caused a significant depression in the alveolar clearance of a test aerosol (^{239}PuO$_2$).

8.2 Short- and long-term exposures

8.2.1 Short-term exposure

8.2.1.1 Oral

By adding large quantities of beryllium carbonate (1250–2000 mg/kg food) to a normal diet, Guyatt et al. (1933) produced rickets (rachitis) in young rats after 24–28 days on the diet. The bone lesions observed were not a direct effect of beryllium itself, but were due to intestinal precipitation of beryllium phosphate and concomitant phosphorus deprivation.

8.2.1.2 Inhalation

Acute chemical pneumonitis occurred in various animal species following the inhalation of beryllium metal or different beryllium compounds (Stokinger, 1981). Repeated, daily, 6-h exposures to beryllium sulfate mist (mean concentration 47 mg BeSO$_4$.6H$_2$O/m^3 or 2 mg Be/m^3) were lethal for goats (100% deaths), guinea-pigs (60%), monkeys (100%), rats (90%), dogs (80%), cats (80%), rabbits (10%), hamsters (50%), and mice (10%). Death occurred within the first week in the goats, guinea-pigs, and monkeys, in the second or third week in the rats, and after one or two months in the dogs and cats. Exposure to 0.95 mg BeSO$_4$.6H$_2$O/m^3 (0.04 mg Be/m^3), over 100 days, did not lead to any deaths in the species tested (Stokinger et al., 1950a). Beryllium fluoride was more toxic than the sulfate. Inhalation of 1 mg BeF$_2$/m^3 (0.19 mg Be/m^3)

caused lung lesions in cats, dogs, rabbits, and rats that were similar to those produced by 10 mg $BeSO_4.6H_2O/m^3$ (0.42 mg Be/m^3). The lesions in the lungs closely resembled those in man, but were not identical. They were most severe and extensive in dogs, minimal in rats, and intermediate in cats and rabbits (Stokinger et al., 1950b, 1953).

In addition to pulmonary injuries, a macrocytic anaemia was observed in dogs and rabbits, with a tendency in dogs to return to a normal blood pattern, despite continued exposure (Stokinger et al., 1950a).

To compare the biological action of highly soluble beryllium fluoride (BeF_2) and beryllium sulfate ($BeSO_4$) with that of the poorly soluble beryllium phosphate ($BeHPO_4$), Schepers (1964) exposed monkeys, 4 in each group, to aerosols of these compounds for 7–30 days, at a concentration of approximately 200 µg Be/m^3. BeF_2 proved most toxic and $BeHPO_4$ least toxic. The initial response was a marked anorexia in the BeF_2 group, and a moderate to slight loss of appetite in the 2 other exposure groups. Dyspnoea, a typical sign of human chronic beryllium disease, constituted another striking sign, particularly in the animals exposed to BeF_2. Moderate recovery was noted, though, after cessation of exposure, some animals died. In another study, mortality was 100% in monkeys exposed to high beryllium phosphate concentrations of around 1140 and 8380 µg Be/m^3.

The histological picture resembled that in other experimental animals and in human beings. Pulmonary oedema and congestion and marked changes in the liver, kidneys, adrenals, pancreas, thyroid, and spleen were found in the animals exposed to BeF_2 and the higher concentrations of $BeHPO_4$. Granulomas were noted in some monkeys, but, unlike the sarcoidotic lesions characteristic for man, they were of simple composition and confined to local lesions within the alveolar walls (Schepers, 1964).

Insoluble beryllium compounds can also produce acute pneumonitis. Hall et al. (1950) found that the toxicity of inhaled BeO depended on the physical and chemical properties of the compound, notably ultimate particle size, state of aggregation, and solubility, which, in turn, were governed by the production conditions. Two high-fired BeO grades (1350 °C and 1150 °C) did not produce pulmonary damage in dogs, guinea-pigs, cats, or rats, exposed for up to 360 h to concentrations of up to 88 mg BeO/m^3 (32 mg Be/m^3). However, exposure to low-fired BeO (400 °C), at 10 mg BeO/m^3 (3.6 mg Be/m^3) for 40 days, caused mortality in rats and marked lung damage in dogs, apparently because of its smaller ultimate particle size and lesser degree of aggregation compared with the high-fired grades. Crossmon & Vandemark (1954) and Spencer et al. (1968) confirmed these results.

8.2.1.3 Other

Cloudman et al. (1949) observed symptoms of osteosclerosis in the form of irregular thickening of the cortices of the long bones, pelvis, and skull of rabbits, 94 days after repeated intravenous injections of a total of 17 mg Be as zinc beryllium silicate.

8.2.2 Long-term exposure

8.2.2.1 Oral

The low bioavailability and, hence, toxicity of ingested beryllium was confirmed by Schroeder & Mitchener (1975 a,b). Exposure of rats and mice to beryllium, in the form of beryllium sulfate in the drinking-water (5 mg Be/litre), did not show any effects on life span and survival. Only slight effects on the body weight of female mice occurred. A mild weight depression was also the only response observed in rats receiving the highest dietary concentration (500 mg BeSO$_4$/kg) in a 2-year feeding study (Morgareidge et al., 1977).

8.2.2.2 Inhalation

The characteristic non-malignant response to long-term, low-level, inhalation exposure to soluble and insoluble beryllium compounds is chronic pneumonitis associated with granulomas (Stokinger, 1981). However, as with short-term exposure, the beryllium granulomatosis observed in experimental animals only partly corresponds with the chronic disease in human beings.

In the early studies of Policard (1948; 1949a; 1949b; 1950a; 1950b), temporary nodular granulomas were observed in the lungs of guinea-pigs exposed to dusts of BeO, $NaBeF_3$, $Be(OH)_2$, beryl, or elemental beryllium. Since inhalation of pure beryllium or beryl dust only produced a temporary pulmonary reaction without formation of granulomas, the fluoride (NaF), which was present in the beryllium compounds administered, was believed to be the causative agent. However, later studies established that beryllium itself caused the pulmonary changes.

Schepers et al. (1957) exposed rats to a $BeSO_4$ aerosol (Table 11) daily, for up to 6 months, and found that the characteristic response was a stimulation of epithelial cell proliferation without any formation of fibrotic tissue. Six main pulmonary reactions occurred during the 6 months following cessation of exposure: formation and clustering of foamy macrophages; infiltration of the alveolar walls; lobular septal cell proliferation; epithelialization of the peribronchial alveolar walls; granulomatosis with a central core of large macrophages and a superficial thin zone of plasma cells; and neoplasia. A similar picture was seen in various long-term studies summarized in Table 11.

In a study by Wagner et al. (1969), the inhalation toxicities of beryl and bertandite aerosols were compared at the "nuisance limit" for all dusts (15 mg/m^3) in rats, hamsters, and monkeys. At this particle concentration, the beryllium contents of the aerosols were 620 and 210 $\mu g/m^3$ for beryl and bertrandite, respectively. Exposure was continued intermittently for 17 months. Pulmonary neoplasia was produced in the beryl-exposed rats, whereas, in the bertrandite-exposed rats and in all other species, the lesions were characterized as atypical proliferation and/or granulomatous inflammation. The importance of genetic predisposition was demonstrated by Barna et al. (1984) who administered single doses of 10 mg BeO intratracheally to 2 commercial strains of guinea-pigs. Granulomatous lung disease occurred in all the animals in one strain, but not in the other.

8.3 Skin irritation and sensitization

Dutra (1951) implanted beryllium compounds and metallic beryllium in the subcutaneous tissues at 12 sites on the side of a pig. Beryllium granulomas similar to those observed in human beings were produced.

Table 11. Non-malignant pulmonary changes in experimental animals after long-term inhalation exposure to beryllium compounds

Species	Beryllium compound	Concentration (μg Be/m^3)	Maximum exposure	Non-malignant pulmonary changes	Reference
Rat	BeSO$_4$	36	26 weeks (8 h/day; 5.5 day/week)	Foam-cell clusters; focal mural infiltration; lobular septal cell proliferation; peribronchial alveolar wall epithelialization; granulomatosis	Schepers et al. (1957)
Rat	BeSO$_4$	2.8 21 42 194	80 weeks (7 h/day; 5 day/week)	No specific inflammatory abnormalities Inflammatory changes in long-surviving rats Chronic pneumonitis; focal granulomatous lesions Acute disease	Vorwald et al. (1966)
Rat	BeSO$_4$	34	72 weeks (7 h/day; 5 day/week)	Increase in lung weight; inflammatory and proliferative changes; clusters of macrophages in alveolar spaces; occasionally granulomatosis and fibrosis	Reeves et al. (1967)
Rat	Bertrandite	210	73 weeks (6 h/day; 5 day/week)	Granulomatous lesions; lymph nodes composed of dust-laden macrophages	Wagner et al. (1969)
Rat	Beryl	620	73 weeks (6 h/day; 5 day/week)	Atypical cell proliferations; loose collections of foamy macrophages; no granulomatous lesions	Wagner et al. (1969)

Table 11 (continued).

Species	Beryllium compound	Concentration (μg Be/m³)	Maximum exposure	Non-malignant pulmonary changes	Reference
Hamster	Bertrandite	210	73 weeks (6 h/day; 5 day/week)	Few granulomatous lesions; atypical proliferations	Wagner et al. (1969)
Hamster	Beryl	620	73 weeks (6 h/day; 5 day/week)	No granulomatous lesions; atypical cell proliferations	Wagner et al. (1969)
Monkey	BeSO$_4$	35	unknown	Pneumonitis; granulomatosis	Vorwald et al. (1966)
Monkey	Bertrandite	210	99 weeks (6 h/day; 5 day/week)	Clusters of dust-loaden macrophages, lymphocytes and plasma cells; no other marked changes	Wagner et al. (1969)
Monkey	Beryl	620	99 weeks (6 h/day; 5 day/week)	Clusters of dust-loaden macrophages, lymphocytes and plasma cells; no other marked changes	Wagner et al. (1969)
Monkey	BeO (1400 °C)	3300–4400	30 min/month (3 times)	No major changes	Conradi et al. (1971)
Dog	BeO (1400 °C)	3300–4400	30 min/month (3 times)	No major changes	Conradi et al. (1971)

Cutaneous hypersensitivity occurred in guinea-pigs after intradermal injection of soluble beryllium compounds (Alekseeva, 1966). These findings were confirmed by other studies decribed in section 8.7.2.

8.4 Reproduction, embryotoxicity, and teratogenicity

Puzanova et al. (1978) injected $BeCl_2$ subgerminally in chicken embryos at early stages of embryogenesis. Doses exceeding 0.3 µg were embryo-lethal; doses of 0.03–0.3 µg were not lethal, but caused malformations consisting of malposed and malformed heart and caudal regression, body wall aplasia, symptoms of the strait jacket syndrome, and malformations of the mandible and facial clefts.

In a study on the transplacental absorption of beryllium (section 6.2), Bencko et al. (1979a) found that beryllium penetration through the placenta was relatively poor, depending on the time of administration. After intravenous administration of 0.1 mg $BeCl_2$/kg body weight on the 14th day of pregnancy, mouse fetuses contained an average of 0.0013 mg Be/kg, which was about one order of magnitude higher than the relative beryllium content of 0.0002 mg/kg, found after administration on the 7th day of pregnancy.

Tsujii & Hoshishima (1979) studied the behaviour of the offspring of mice exposed to beryllium sulfate during pregnancy. Six female CFW strain mice received 11 intraperitoneal injections of $BeSO_4$ (140 ng Be/mouse per day) during pregancy. The behaviour of the offspring differed from that of the control animals as follows: delayed response in head turning in a geotaxis test, acceleration in a straight-walking test, delayed bar-holding response, acceleration of bar holding.

The effects of beryllium nitrate on early and late pregnancy in rats were investigated by Mathur et al. (1987). When beryllium was injected intravenously at a dose of 0.316 mg/kg body weight on day 1 of pregnancy, implantation and late pregnancy phase were not affected. Pups, which appeared to be normal when delivered, died after 2–3 days, probably due to the toxic effects of beryllium. When beryllium was administered on day 11 following mating, all the fetuses were resorbed because of the immediate entry of beryllium into the fetal circulation. Administration of beryllium after the formation of the placenta, i.e., after day 12, prevented the fetal resorption.

8.5 Mutagenicity and related end-points

8.5.1 DNA damage

Kubinski et al. (1977,1981) reported that salts of beryllium (30 μmol/litre) induced complexes between DNA and proteins in *Escherichia coli* cells and Ehrlich ascites cells, suggesting an interaction of beryllium with DNA.

The beryllium ion also seems to bind to DNA polymerases, since beryllium chloride caused a dose-related (2–10 mmol/litre) reduction in the accuracy of DNA synthesis in an *in vitro* assay using purified DNA polymerase from avian myeloblastosis virus (Sirover & Loeb, 1976).

Divalent beryllium was also found to increase misincorporation of nucleoside triphosphates during polymerization of poly-d (A-T) by *Micrococcus luteus* DNA polymerase (Luke et al., 1975). Associated with this effect was a strong inhibition of one of the two exonuclease activities of this enzyme. An increase in misincorporation was also reported by Zakour et al. (1981).

In the *E. coli* pol$^+$/pol$^-$ assay for DNA-modifying effects, beryllium sulfate was inactive, with or without an S-9 activation system (Rosenkranz & Poirier, 1979).

In the hepatocyte primary culture/DNA repair test, beryllium sulfate tetrahydrate (0.1–1 g /litre) was negative (Williams et al., 1982).

8.5.2 Mutation

8.5.2.1 Bacteria and yeast

Beryllium sulfate was not mutagenic in several bacterial mutation assays. In the *Salmonella typhimurium* test, frequency of the backward mutations was not significantly enhanced in the most commonly used strains (TA 1535, TA 1536, TA 1537, TA 98, TA 100) (Rosenkranz & Poirier, 1979; Simmon, 1979a).

Simmon et al. (1979) tested beryllium sulfate in the host-mediated assay with different *Salmonella* strains (TA 1530, TA 1535, TA 1538) and *Saccharomyces cerevisiae* D3. Beryllium sulfate given to mice either orally (1200 mg/kg) or by intramuscular injection (25 mg/kg) was not mutagenic.

Beryllium chloride (10 mmol/litre) produced a weak mutagenic response in the *Bacillus subtilis* rec. assay (Kanematsu et al., 1980). However, no effect was noted in another rec. assay using 50 mmol beryllium chloride/litre (Nishioka, 1975). McCann et al. (1975) hypothesized that the large amounts of magnesium salts, citrate, and phosphate in the minimal medium of bacterial tests may preclude the entry of beryllium into bacteria. The negative mutagenic response of beryllium sulfate in the mitotic recombination assay with *Saccharomyces cerevisiae* strain D3 (Simmon, 1979b) may also be due to reduced penetration of beryllium into the cells.

8.5.2.2 Cultured mammalian cells

Beryllium caused gene mutations in cultured mammalian cells. At concentrations of 2 and 3 mmol/litre, beryllium chloride enhanced the induction of 8-azaguanine-resistant mutants in the Chinese hamster V79 cells 6-fold compared with control values. The underlying mutation resulted in a deficiency in the activity of the enzyme hypoxanthine guanine phosphoribosyl transferase (Miyaki et al., 1979). Hsie et al. (1979a,b) reported similar results using beryllium sulfate.

8.5.3 Chromosomal effects

Beryllium caused marked chromosomal aberrations in cultured mammalian cells. Talluri & Guiggiani (1967) reported that beryllium chloride (0.5–10 mmol/litre) caused stickiness, chromatid gaps and breaks, fragments, and mitotic delay in cultured peripheral lymphocytes and primary kidney cells of the domestic pig.

Beryllium sulfate (0.03 mmol/litre) was clastogenic (i.e., chromosome-breaking) in Syrian hamster embryo cells with 19% aberrations in the treated cells compared with 1.5% in the controls (Larramendy et al., 1981). In the same study, a clear clastogenic potential of beryllium in human lymphocytes was noted, though it was less marked than in the animal cells.

Paton & Allison (1972) did not find any chromosomal aberrations with 1×10^{-5} and 1×10^{-3} mmol beryllium sulfate/litre in human diploid fibroblasts and human leukocytes *in vitro*.

Larramendy et al. (1981) reported a dose-related increase in sister chromatid exchanges in both Syrian hamster embryo cells and human lymphocytes. The exchange frequency in the human lymphocytes was 11.30 ± 0.60, 17.75 ± 1.10, 18.15 ± 1.79, and 20.7 ± 1.01 at $BeSO_4$ concentrations of 0 μg/ml (control), 1 μg/ml, 2.5 μg/ml, and 5 μg/ml culture medium. Similar data were found for the animal cells.

There are no *in vivo* data on the clastogenic potential of beryllium.

8.6 Carcinogenicity

8.6.1 Bone cancer

Gardner & Heslington (1946) investigated the carcinogenic properties of beryllium. Osteosarcomas of the long bones developed in 7 rabbits that survived 7 or more months after the intravenous injection of zinc beryllium silicate ($ZnBeSiO_3$) (which is used in the fluorescent light tube industry). Both zinc oxide and zinc silicate alone were inactive, while beryllium oxide (firing temperature unstated) was also carcinogenic (Table 12).

Several investigators reproduced beryllium bone sarcoma in the rabbit with beryllium metal and various beryllium compounds (Table 12). Intraveneous injection or intramedullary injection, in which beryllium was directly introduced into the medullary cavity of bones, were used in most of these studies.

In one study, osteosarcomas were found in mice after 20–22, twice-weekly, intravenous injections of zinc beryllium silicate; beryllium oxide did not induce any effects (Cloudman et al., 1949). The numbers of mice and incidence of tumours were not stated.

Guinea-pigs and rats did not develop bone cancer after intravenous injection of both zinc beryllium silicate and beryllium oxide (Gardner & Heslington, 1946).

As shown in the summary table (Table 12), the incidence of tumours was consistently high in the rabbit studies, varying from 13 to 100%. The latent period varied from 5.5 to 24 months after the last injection of beryllium (Groth, 1980).

The osteosarcomas developed in different bones, including the humerus, tibia, femur, ilium, ischial tuberosity, lumbar vertebra, scapula, and ribs. Frequently, two or more bones were affected in the same animal. Metastases occurred in 40–100% of the animals, most frequently in the lungs, but also in the liver, kidney, omentum, skin, and lymph nodes (Groth, 1980).

The cell types of the bone tumours were described as osteoblastic, chondroblastic, and fibroblastic, and differed from animal to animal, with all 3 cell types occurring frequently in the same tumour (Kelly et al., 1961). Tapp (1969) characterized the sarcomas as chondrosarcomatous or anaplastic. The metastases appeared to be similar in all respects to the primary tumours and contained osseous tissue (Barnes et al., 1950; Dutra & Largent, 1950; Dutra et al., 1951). The metastatic osteogenic sarcomas developing in the lungs could easily be distinguished from primary lung neoplasms.

8.6.2 Lung cancer

Vorwald (1953) first reported experimental evidence for pulmonary tumours induced by beryllium. These observations were confirmed in several studies on rats following the inhalation of beryllium sulfate, oxide, phosphate, fluoride, or beryl ore. Positive results were also seen after intratracheal injection of beryllium sulfate, oxide, metal, and various beryllium alloys. Pulmonary carcinomas were produced in monkeys following inhalation of beryllium sulfate and phosphate and after intrabronchial implantation of a beryllium oxide suspension. No pulmonary neoplasms were found in carcinogenicity studies on rabbits, guinea-pigs, or hamsters. The results of the various inhalation and intratracheal/intrabronchial studies are summarized in Tables 13 and 14.

Table 12. Osteosarcoma from beryllium[a]

Compound[b]	Species	Total dose (mg Be)	Mode of administration[c]	Incidence of tumours	Incidence of metastases	Reference
ZnBeSiO$_x$	rabbit	60	iv in 20 doses	7/7 (100%)	3/7 (43%)	Gardner & Heslington (1946)
BeO	rabbit	360	iv in 20 doses	unknown	unknown	
ZnBeSiO$_x$	guinea-pig	60	iv in 20 doses	0	-	
BeO	guinea-pig	360	iv in 20 doses	0	-	
ZnBeSiO$_x$	rat	60	iv in 20 doses	0	-	
BeO	rat	360	iv in 20 doses	0	-	
ZnBeSiO$_x$	rabbit	17	iv in 20-22 doses	4/5 (80%)	3/4 (75%)	Cloudman et al. (1949)
BeO	rabbit	140	iv in 20-22 doses	0	-	
ZnBeSiO$_x$	mouse	0.26	iv in 20-22 doses	unknown	unknown	
BeO	mouse	0.55	iv in 20-22 doses	0	-	
ZnMnBeSiO$_x$	rabbit	3.7-7.0	iv in 1-30 doses	3/6 (50%)		Hoagland et al. (1950)
ZnMnBeSiO$_x$	rabbit	10-12.6	iv in 1-30 doses	3/4 (75%)	5/7 (71%)	
BeO	rabbit	360	iv in 1-30 doses	1/9 (11%)		
ZnBeSiO$_x$	rabbit	7.2	iv in 6 doses	4/14 (29%)	2/4 (50%)	Barnes et al. (1950)
ZnBeSiO$_x$	rabbit	15	iv in 10 doses	2/3 (67%)	1/2 (50%)	
BeSiO$_x$	rabbit	180	iv in 6-10 doses	1/11 (9%)	0/1	
ZnBeSiO$_x$	rabbit	64-90	iv in 17-25 doses	2/3 (67%)	2/2 (100%)	Dutra & Largent (1950)
BeO	rabbit	360-700	iv in 20-26 doses	6/6 (100%)	6/6 (100%)	
ZnBeSiO$_x$	rabbit	12	iv in 20 doses	5/10 (50%)	>2/5 (40%)	Janes et al. (1954)

Table 12. (continued)

ZnBeSiO$_x$	rabbit	12	iv in 20 doses	10/13 (77%)	unknown	Kelly et al., (1961)
BeO	rabbit	360	iv	unknown	2/3 (66%)	Komitowski (1967)
BeO	rabbit	79-144	IMD	7/9 (78%)	unknown	Yamaguchi (1963)
BeO	rabbit	151-216	IMD	11/11 (100%)	unknown	
ZnBeSiO$_3$	rabbit	0.144	IMD	4/12 (33%)	3/4 (75%)	Tapp (1969)
BeO	rabbit	1 mg/m^3	inhalation:	0/5	-	Dutra et al. (1951)
		6 mg/m^3	25 h/week, 9-13	1/6	1/1	
		30 mg/m^3	months	0/8	-	

[a] Adapted from: Reeves (1978) and Groth (1980).
[b] ZnBeSiO$_x$ = zinc beryllium silicate;
ZnMnBeSiO$_x$ = zinc manganese beryllium silicate;
BeO = beryllium oxide;
BeSiO$_x$ = beryllium silicate.
[c] IMD = intramedullary; iv = intravenous.

Table 13. Pulmonary cancer after inhalation exposure to beryllium[a]

Species	Compound	Atmospheric concentration (Be)	Duration of exposure	Incidence of pulmonary carcinomas	Reference[d]
Rat	BeSO$_4$	33-35 μg/m^3	12-14 months, 33-38 h/week	4/8	Vorwald (1953)*
Rat	BeSO$_4$	33-35 μg/m^3	13-18 months, 33-38 h/week	17/17	Vorwald et al. (1955)**
Rat	BeSO$_4$	55 μg/m^3	3-18 months, 33-38 h/week	55/74	Vorwald (1962)*
Rat	BeSO$_4$	180 μg/m^3	12 months, 33-38 h/week	11/27	Vorwald (1962)*
Rat	BeSO$_4$	18 μg/m^3	3-22 months, 33-38 h/week	72/103	Vorwald (1962)*
Rat	BeSO$_4$	18 μg/m^3	8-21 months, 33-38 h/week	31/63	Vorwald (1962)*
Rat	BeSO$_4$	18 μg/m^3	9-24 months, 33-38 h/week	47/90	Vorwald (1962)*
Rat	BeSO$_4$	18 μg/m^3	11-16 months, 33-38 h/week	9/21	Vorwald (1962)*
Rat	BeSO$_4$	1.8-2.0 μg/m^3	8-21 months, 33-38 h/week	25/50	Vorwald (1962)*
Rat	BeSO$_4$	1.8-2.0 μg/m^3	9-24 months, 33-38 h/week	43/95	Vorwald (1962)*
Rat	BeSO$_4$	1.8-2.0 μg/m^3	13-16 months, 33-38 h/week	3/15	Vorwald (1962)*
Rat	BeO	9000 μg/m^3	3-12 months, 33-38 h/week	22/36	Vorwald (1962)*
Rat	BeSO$_4$	21-42 μg/m^3	18 months, 33-38 h/week	almost 100%[b]	Vorwald et al. (1966)*
Rat	BeSO$_4$	2.8 μg/m^3	18 months, 33-38 h/week	13/21	Vorwald et al. (1966)*
Rat	BeSO$_4$	32-35 μg/m^3	6-9 months, 44 h/week	58/136[c]	Schepers et al. (1957)**
Rat	BeHPO$_4$	32-35 μg/m^3	1-12 months	35-60/170	Schepers (1961)*
Rat	BeHPO$_4$	227 μg/m^3	1-12 months	7/40	Schepers (1961)*
Rat	BeF$_2$	9 μg/m^3	6-15 months	10-12/200	Schepers (1961)*
Rat	ZnMnBeSiO$_3$	850-1250 μg/m^3	1-9 months	4-20/220	Schepers (1961)*
Rat	BeSO$_4$	34 μg/m^3	13 months, 35 h/week	43/43	Reeves et al. (1967)
Rat	BeSO$_4$	36 μg/m^3	3 months, 35 h/week	19/22	Reeves & Deitch (1969)
Rat	BeSO$_4$	36 μg/m^3	6 months, 35 h/week	33/33	Reeves & Deitch (1969)

Table 13 (continued)

Species	Compound	Concentration	Duration	Result	Reference
Rat	BeSO$_4$	36 µg/m^3	9 months, 35 h/week	15/15	Reeves & Deitch (1969)
Rat	BeSO$_4$	36 µg/m^3	12 months, 35 h/week	21/21	Reeves & Deitch (1969)
Rat	BeSO$_4$	36 µg/m^3	18 months, 35 h/week	13/15	Reeves & Deitch (1969)
Rat	BeO	400 µg/m^3	4 months, 5 h/week	8/21	Litvinov et al. (1984)
Rat	BeO	30 µg/m^3	4 months, 5 h/week	6/26	Litvinov et al. (1984)
Rat	BeO	4 µg/m^3	4 months, 5 h/week	4/39	Litvinov et al. (1984)
Rat	BeO	0.8 µg/m^3	4 months, 5 h/week	3/44	Litvinov et al. (1984)
Rat	BeCl$_2$	400 µg/m^3	4 months, 5 h/week	11/19	Litvinov et al. (1984)
Rat	BeCl$_2$	30 µg/m^3	4 months, 5 h/week	8/24	Litvinov et al. (1984)
Rat	BeCl$_2$	4 µg/m^3	4 months, 5 h/week	2/42	Litvinov et al. (1984)
Rat	BeCl$_2$	0.8 µg/m^3	4 months, 5 h/week	1/41	Litvinov et al. (1984)
Rat	beryl	620 µg/m^3	17 months, 30 h/week	18/19	Wagner et al. (1969)
Rat	bertrandite	210 µg/m^3	17 months, 30 h/week	0/30-60	Wagner et al. (1969)
Rabbit	ZnMnBeSiO$_3$	1000 µg/m^3	24 months	0	Schepers (1961)
Guinea pig	ZnMnBeSiO$_3$	1000 µg/m^3	22 months	0	Schepers (1961)
Guinea pig	BeSO$_4$	35 µg/m^3	12 months	0	Schepers (1961)
Guinea pig	BeSO$_4$	3.7-30.4 µg/m^3	18-24 months, 30 h/week	0/58	Reeves et al. (1972)
Hamster	beryl	620 µg/m^3	17 months, 30 h/week	0/48	Wagner et al. (1969)
Hamster	bertrandite	210 µg/m^3	17 months, 30 h/week	0/48	Wagner et al. (1969)
Monkey	BeSO$_4$	38.8 µg/m^3	>36 months, 15 h/week	8/11	Vorwald (1968)
Monkey	BeSO$_4$	35-200 µg/m^3	8 days, 6 h/day	0/4	Schepers (1964)

Table 13 (continued)

Species	Compound	Atmospheric concentration (Be)	Duration of exposure	Incidence of pulmonary carcinomas	Reference[d]
Monkey	BeF$_2$	180 µg/m^3	8 days, 6 h/day	0/4	Schepers (1964)
Monkey	BeHPO$_4$	200 µg/m^3	8 days, 6 h/day	0/4	Schepers (1964)
Monkey	BeHPO$_4$	1100 µg/m^3	8 days, 6 h/day	1/4	Schepers (1964)
Monkey	BeHPO$_4$	8300 µg/m^3	8 days, 6 h/day	0/4	Schepers (1964)
Monkey	beryl	620 µg/m^3	17 months, 30 h/week	0/12	Wagner et al. (1969)
Monkey	bertrandite	210 µg/m^3	17 months, 30 h/week	0/12	Wagner et al. (1969)

[a] From Reeves (1978), adapted and supplemented.

[b] Number of animals not stated.

[c] Number of tumour-bearing animals not stated; total number of tumours: 76.

[d] These studies were not published as primary experimental publications, but were either quoted in reviews (*) or published as abstracts (**). The documentation of experimental details, including verification of chamber exposure concentration values, is unavailable. These figures, therefore, must be treated with caution.

Table 14. Pulmonary cancer after intratracheal or intrabronchial injection of beryllium

Species	Compound	Total dose (Be)	Mode of administration	Autopsy interval[a]	Incidence of pulmonary carcinomas	Reference
Rat	ZnMnBeSiO$_x$	0.46 mg	intratracheal	ns	0	Vorwald (1950)
Rat	BeO	0.338 mg	intratracheal in 3 doses	ns	1/4	Vorwald (1953)
Rat	BeSO4	0.033 mg	intratracheal in 3 doses	ns	1/5	Vorwald (1953)
Rat	BeO (500 °C)	9 mg	intratracheal	30-77 weeks	23/45	Spencer et al. (1968)
Rat	BeO (1100 °C)	9 mg	intratracheal	30-69 weeks	3/19	Spencer et al. (1968)
Rat	BeO (1600 °C)	9 mg	intratracheal	32-97 weeks	3/28	Spencer et al. (1968)
Rat	Be metal	2.5 mg	intratracheal	16-19 months	6/6	Groth et al. (1980)
Rat	Be metal	0.5 mg	intratracheal	16-19 months	2/3	Groth et al. (1980)
Rat	Be metal, passivated	2.5 mg	intratracheal	16-19 months	4/4	Groth et al. (1980)
Rat	Be metal, passivated	0.5 mg	intratracheal	16-19 months	7/11	Groth et al. (1980)
Rat	BeAl alloy	1.55 mg	intratracheal	16-19 months	2/6	Groth et al. (1980)
Rat	BeAl alloy	0.3 mg	intratracheal	16-19 months	1/9	Groth et al. (1980)
Rat	4% BeCu alloy	0.1 mg	intratracheal	16-19 months	0/11	Groth et al. (1980)
Rat	2.2% BeNi alloy	0.056 mg	intratracheal	16-19 months	0/12	Groth et al. (1980)
Rat	2.4% BeCuCo alloy	0.06 mg	intratracheal	16-19 months	0/15	Groth et al. (1980)
Rat	BeO (900 °C)	15 mg	intratracheal in 15 doses	18 months	7/29	Ishinishi et al. (1980)
Rabbit	ZnMnBeSiO$_x$	2.3-6.9 mg	intratracheal	ns	0	Vorwald (1950)

Table 14 (continued)

Species	Compound	Total dose (Be)	Mode of administration	Autopsy interval[a]	Incidence of pulmonary carcinomas	Reference
Guinea-pig	ZnMnBeSiO$_x$	3.4 mg	intratracheal	ns	0	Vorwald (1950)
Guinea-pig	Be stearate	5 mg	intratracheal	ns	0	Vorwald (1950)
Guinea-pig	Be(OH)$_2$	31 mg	intratracheal	ns	0	Vorwald (1950)
Guinea-pig	Be metal	54 mg	intratracheal	ns	0	Vorwald (1950)
Guinea-pig	BeO	75 mg	intratracheal	ns	0	Vorwald (1950)
Monkey	BeO	18-90 mg	bronchomural implant + intrabronchial injection	ns	3/20	Vorwald (1968)

[a] ns = not specified.

The tumours observed in rats were usually adenocarcinomas (Vorwald, 1953; Schepers et al., 1957; Reeves et al., 1967). Some of the tumours developed metastases to tracheobronchial lymph nodes and pleura (Vorwald, 1953) and to the adrenals, kidneys, liver, pancreas, and brain (Schepers, 1961).

While bertrandite did not produce cancer, beryl ore caused pulmonary adenomas, adenocarcinomas, and epidermoid carcinomas in rats (Wagner et al., 1969). This difference is not explained by the lower beryllium concentration in the bertrandite-exposure group, since much lower absolute beryllium concentrations showed positive results (Table 13).

The physical and chemical properties of compounds of beryllium apparently determine their carcinogenic potential, as demonstrated by Spencer et al. (1968). Three different samples of beryllium oxide, calcined at 500, 1100, or 1600 °C, respectively, were injected intratracheally into rats. The results of all 3 studies were positive, but the incidence of pulmonary adenocarcinomas was highest after treatment with the low-fired BeO, where 23/45 or 51% of the rats developed carcinomas, compared with 3/19 or 16% and 3/28 or 11% with the high-fired oxides. In a later study (Spencer et al., 1972), BeO rocket exhaust product proved almost as carcinogenic as low-fired BeO.

Lung tumours were also produced in rats by intratracheal injection of beryllium metal, passivated beryllium metal, and a beryllium-aluminium alloy (62% beryllium) at doses of 0.3–2.5 mg Be/animal (Groth et al., 1980). No tumours were seen after administration of other beryllium alloys. However, the latter were injected in much lower doses (Table 14), possibly obscuring a carcinogenic potential of these compounds.

In a short-term inhalation study carried out by Schepers (1964), a small pulmonary neoplasm was found in 1 out of 20 rhesus monkeys exposed to 1.1 mg Be/m^3 (BeHPO$_4$) for 8 days and autopsied 82 days later. BeSO$_4$ and BeF$_2$ were negative in the short-term, but after long-term exposure to 39 μg Be/m^3, 8 out of 9 monkeys that had survived 6 or more years showed pulmonary tumours of various histological types, all metastazing to the mediastinal lymph nodes and some to the bone, adrenals, and liver (Vorwald, 1968).

From the data in Table 13 and Table 14 it appears that the induction of pulmonary cancer by beryllium is species-specific. While rats and, perhaps, monkeys are susceptible, no pulmonary tumours were observed in rabbits, hamsters, and guinea-pigs. The latter were exposed to concentrations that were carcinogenic in 100% of exposed rats. The reasons for this negative neoplastic response are not known. Whether the cutaneous hypersensitivity to beryllium in guinea-pigs indicates that some form of cellular immunity may be a factor in determining the carcinogenic response to beryllium, remains unresolved (Reeves, 1978). The increased incidence of bone sarcomas in splenectomized rabbits exposed to beryllium (Janes et al., 1954; 1956) seems to indicate the relevance of immunocompetence for the induction of beryllium cancer.

8.7 Mechanisms of toxicity, mode of action

8.7.1 Effects on enzymes and proteins

Beryllium is a potent inhibitor of various enzymes of phosphate metabolism, particularly of alkaline phosphatase. DuBois et al. (1949) found that toxic effects of beryllium might involve interference with the biological functions of magnesium. A 50% inhibition of magnesium-activated phosphatase activity in rat serum was seen at 1.8×10^{-6} mol/litre. Addition of Mg^{2+} did not influence the inhibitory action of beryllium, indicating that beryllium has a much greater affinity for the enzyme than magnesium. Only at very high magnesium concentrations, i.e., at a magnesium/beryllium ratio of 40 000:1, was a decrease in the beryllium-induced inhibition of the alkaline phosphatase activity in rabbit kidney noted (Aldridge, 1950).

Thomas & Aldridge (1966) found that, of various enzymes examined, only alkaline phosphatase and phosphoglucomutase activities were inhibited by 10^{-6} mol $BeSO_4$/litre, whereas the activities of acid phosphatase, phosphoprotein phosphatase, adenosine triphosphatase, glucose-6-phosphatase, polysaccharide phosphorylase, hexokinase, phosphoglyceromutase, ribonuclease, A-esterase, cholinesterase, and chymotrypsin were not inhibited at 10^{-3} mol $BeSO_4$/litre. With phosphoglucomutase, inhibition was competitive with respect to magnesium. However, once established, reversion of the inhibition could not be produced by adding magnesium. Beryllium probably combines with the unphosphorylated enzymes, both alkaline phosphatase and phosphoglucomutase, thus, interfering with the competition of magnesium for the unphosphorylated enzyme.

Cummings et al. (1982) found that cytoplasmic and nuclear cyclic AMP-independent casein kinase I was inhibited by beryllium (BeSO$_4$/sulfosalicylic acid (1:1) 10 μmol/litre) indicating that the phosphorylation of protein substrates is also inhibited by beryllium. Possibly the impairment of key protein phosphorylation is the biochemical basis for many of the toxic and carcinogenic actions of beryllium (Skilleter, 1984).

A number of other enzymes are inhibited by beryllium, but usually at higher concentrations. The action of beryllium on these enzymes might be through its combination with the substrate, depleting the enzyme of its usual magnesium-substrate complex, rather than a direct action on the enzymes (Thomas & Aldridge, 1966).

Beryllium inhibited the magnesium-dependent phosphatic acid phosphatase by 30% at 10^{-4} mol/litre and completely at 10^{-3} mol/litre (Hokin et al., 1963). Alkaline phosphatase was inhibited by 50% at 10^{-3}–10^{-6} mol/litre and adenosine triphosphatase, by 22–35% at 10^{-3} mol/litre (Cochran et al., 1951), deoxythymidine kinase was inhibited by 50% at 10^{-4} mol/litre (Mainigi & Bresnick, 1969); and lactate dehydrogenase was inhibited by 65% at 10^{-3} mol/litre (Schormüller & Stan, 1965).

Beryllium also blocked the tricarboxylic cycle by inhibiting the activity of malic, succinic, and α-ketoglutaric dehydrogenases. This occurred in the liver and lungs of rats after intramuscular administration of 0.22 mg Be/kg (BeSO$_4$); addition of MgSO$_4$ decreased the inhibition (Mukhina, 1967).

The induction of drug-metabolizing enzymes in rat liver was also inhibited by beryllium (Witschi & Marchand, 1971). Intravenous injection of 5×10^{-4} mol Be/kg body weight in rats inhibited hepatic induction of acetanilide hydroxylase, aminopyrine demethylase, and tryptophan pyrrolase, indicating that beryllium can interfere with some gene transcription mechanisms.

Beryllium compounds react selectively only with certain proteins (Reiner, 1971), affecting the cellular distribution of the protein. In rats given 33 mg of beryllium by intratracheal injection, the protein contents of the microsomes in lung-tissue cells almost doubled compared with those in control animals, while no changes occurred in the nuclei or mitochondria (Vorwald & Reeves, 1959).

These observations were confirmed by Parker & Stevens (1979) who showed that, of the chromatin proteins in liver nuclei, it was only 6–17% of the non-histones that bound beryllium, while the histones did not have any affinity for the metal ion.

The results of several studies indicate that the target for beryllium toxicity is the cellular DNA. $BeSO_4$ (10^{-3} mol/litre) inhibited cell division in the metaphase (Chèvremont & Firket, 1951). RNA biosynthesis was not affected. Truhaut et al. (1968) found that $BeSO_4$ caused preferential accumulation of radioberyllium in the nuclei of regenerating rat liver and an increase in the sedimentation constant of DNA.

Witschi (1970) noted inhibition of DNA synthesis by beryllium in regenerating rat liver. This was probably a consequence of inhibition of enzymes that play a critical role in DNA replication, and not a result of the direct interaction of beryllium with enzymes, such as thymidine kinase or DNA polymerase. However, Luke et al. (1975) found strong inhibition of DNA polymerase from *Micrococcus luteus*. Be^{2+} increased misincorporation of polydeoxyadenosyl-thymidine during polymerization and this effect was associated with a strong inhibition of the 3'-5' exonuclease activity.

Be^{2+} was the only one of several divalent cations that altered the accuracy of DNA synthesis using purified DNA polymerase from avian myeloblastosis virus (Sirover & Loeb, 1976). It probably does not interact with the catalytically active Mg^{2+} sites on DNA polymerase, but with a non-catalytic site.

Beryllium salts have been shown to exhibit dose-dependent stimulation and inhibition of both murine lymphocyte and accessory cell activities *in vitro* (Skilleter, 1986). In sheep, both insoluble $Be(OH)_2$ and chemically complexed Be caused a powerful immunoblast proliferation (Denham & Hall, 1988; Hall, 1988).

8.7.2 Immunological reactions

Beryllium hypersensitivity appears to be cell-mediated and of the delayed type. Alekseeva (1966) injected 2.5 or 10 μg Be as the chloride intradermally in guinea-pigs that had been sensitized to beryllium 4 weeks earlier. At the higher dose, marked inflammatory reactions were seen at the site of injection after a few days; at the lower dose, repeated injections were necessary to evoke reactions. By transferring homogenates of lymphoid tissue from sensitized to unsensitized animals, the hypersensitivity to beryllium could be transferred. Transferring serum proved negative in this respect. The findings of Alekseeva (1966) were confirmed by other studies. Chiappino et al. (1969) demonstrated that all cutaneous reactions to beryllium in guinea-pigs could be inhibited by injection of an anti-lymphocytic serum from rabbits. Turk & Polak (1969) suppressed reactivity by intravenous injection of beryllium lactate, and Reeves et al. (1972) observed a suppressed cutaneous reactivity in guinea-pigs after inhalation exposure to beryllium. Conversely, the intradermally treated animals developed less severe pulmonary lesions than normal animals.

The mode of administration and the properties of beryllium compounds influence the magnitude of the immunological response. Krivanek & Reeves (1972) sensitized guinea-pigs with the sulfate, the serum albuminate, the hydrogen citrate, and the aurintricarboxylate of beryllium. The two latter complexes, in which beryllium is strongly bound and, thus, unavailable for interaction with the decisive molecule, did not produce any immunological reactions. Surprisingly, serum beryllium albuminate evoked a stronger reaction than $BeSO_4$. It has been assumed that the beryllium ion acts as a hapten, and that the beryllium serum albuminate is either identical with, or very similar to, the complete antigen.

Vacher (1972) found that delayed hypersensitivity in the skin of guinea-pigs resulted only from skin contact with beryllium. Thus, parenteral administration would not elicit immunological reactions. Moreover, only the forms of beryllium that are capable of producing a complex with skin constituents were immunogenic.

9. EFFECTS ON HUMAN BEINGS

9.1 General population exposure

From the use pattern of beryllium it can be deduced that toxicologically relevant exposure to beryllium is largely confined to the workplace. Only a few exposure situations have been reported for the general population, i.e., the use of mantle-type camp lanterns, the handling of broken fluorescent tubes, and the "neighbourhood" cases with indirect exposure outside beryllium-producing or beryllium-processing plants.

First recognized in 1938 by Gelman (1938), "neighbourhood" cases gained considerable interest in the 1940s. Several non-occupational cases in individuals living in the close vicinity of beryllium plants and "para-occupational" cases in beryllium workers' families were reported (section 5.2). By 1966, a total of 60 "neighbourhood" cases had been reported in the USA, 27 of which were related solely to contact with worker's clothes, 18 to air contact alone, and 13 to clothes plus air contact; for 2 cases, no exposure data were available (Hardy et al., 1967). There were at least 3 children among these cases (Hall et al., 1959).

Eleven cases of chronic beryllium disease with symptoms similar to those found in beryllium workers (section 9.2) were diagnosed among residents in the close vicinity of a beryllium production plant in Ohio, USA (Eisenbud et al., 1949; Eisenbud, 1982). In a retrospective investigation, Eisenbud et al. (1949) concluded that 10 out of the 11 non-occupational cases lived within 1.2 km of the plant and that no members of their households had worked in the plant. During a 10-week air-sampling period in 1948, average concentrations of beryllium at a distance of 1.2 km were found to range from 0.004 to 0.02 $\mu g/m^3$. Taking into account the operating history of the plant, it was estimated that beryllium concentrations could have been greater, in the past, by approximately a factor of 10. It was assumed that "the lowest exposure that produced disease was greater than 0.01 $\mu g/m^3$ and probably less than 0.1 $\mu g/m^3$" (Eisenbud, 1982). On the basis of these studies, a maximum ambient air level of 0.01 µg/m³ was recommended and adopted by some regulatory agencies (section 5.2).

One of the 11 cases lived about 2.8 km from the plant, but was exposed to beryllium through the work clothes of her husband, who had worked in the plant for 3 months (Eisenbud et al., 1949). A similar case was reported by Hardy (1948). The mother of a female worker in a fluorescent lamp plant came in contact with beryllium dust from her daughter's shoes and clothes. Both mother and daughter developed chronic beryllium disease, which was fatal for the mother.

In one instance reported by Lieben & Williams (1969), the individuals affected lived far away from the beryllium plant, but had regularly visited a graveyard situated across the street from the beryllium refinery.

Once the potential health hazards of beryllium were recognized and accepted, highly improved emission control and hygiene measures were established in beryllium plants. Hence, no "neighbourhood" cases have been reported in recent years.

Bencko et al. (1980) conducted an epidemiological study on groups of people exposed occupationally (section 9.2.1.2) and non-occupationally (36 persons) to emissions from Czechoslovakian power plants that burned coal with a comparatively high beryllium content. The average beryllium concentration in the ambient air (measured by the fluorometric method) in the vicinity of a power plant was 0.08 µg/m³. Immunological changes, in terms of elevated levels of IgG and IgA, and increased levels of autoantibodies and antibodies against antigens obtained from organs of rats with experimental berylliosis, were found in comparison with a control group of healthy subjects who had no contact with beryllium or other industrial toxic agents. These immune reactions can be considered to be signs of beryllium exposure.

Because of the high sensitization potential of beryllium in provoking contact allergies (section 9.2), the increasing use of beryllium in dentistry (section 3.3) may be important in terms of general population exposure (Bencko, 1989). Schönherr & Pevny (1985) presented 9 cases of a patch test-positive beryllium allergy, of which 5 showed an allergic contact stomatitis, probably caused by beryllium-containing dental prostheses or cement. Other metals proved negative, except for one case of a positive reaction to cobalt.

9.2 Occupational exposure

9.2.1 Effects of short- and long-term exposures

The earliest reports describing a disease in beryllium workers appeared in Europe in the 1930s and early 1940s (Weber & Engelhardt, 1933; Marradi-Fabroni, 1935; Gelman, 1936; Gelman, 1938; Menesini, 1938; Berkovitz & Israel, 1940; Meyer, 1942; Wurm & Ruger, 1942). The clinical symptoms resembled metal fume fever, chemical pneumonitis, and related pulmonary irritations associated with irritant gases, such as phosgene or chlorine, and corrosive acids and alkalis. Hence, the symptoms, which are now generally accepted as being typical for the acute disease caused by beryllium, were erroneously related to the anions in the beryllium compounds, i.e., fluorides and oxyfluorides. Subsequently, Kress & Crispell (1944) and Van Ordstrand et al. (1945) reported the toxic potential of beryllium itself, and Hardy & Tabershaw (1946) found evidence for a chronic beryllium disease.

Following these and other reports, it was soon generally accepted that 2 principal types of disease could be produced by the biological action of beryllium after inhalation and/or dermal exposure, namely acute and chronic beryllium disease. The main differences between these 2 types are summarized in Table 15. In contrast to many other xenobiotics, the duration of exposure does not necessarily govern the type of disease, since low-level exposure of a few hours has been reported to produce a chronic beryllium disease similar to that following years of exposure. Similarly, brief but massive, or prolonged but less intensive, exposure to beryllium may cause the acute disease.

9.2.1.1 Acute disease

Tepper et al (1961) defined acute beryllium disease as follows: "to include those beryllium-induced disease patterns with less than one

Table 15. Classification of beryllium-induced non-neoplastic diseases[a]

Specification of beryllium	Type of exposure	Manifestation of disease	Duration	Clinical form	Degree or stages	Outcome, remote effects and complications
Soluble	acute[b]	Rapid (within 3 days)	< 1 year	Acute beryllium disease: nasopharyngitis, tracheobronchitis, bronchiolitis, pneumonitis, conjunctivitis, dermatosis	light, medium, severe	Pneumosclerosis; chronic bronchitis; bronchiectasia; emphysema, bronchial asthma; pulmonary insufficiency; pulmonary heart; cardio-vascular insufficiency; recovery possible if not fatal
	short-term[c]	Delayed for several weeks				
Poorly soluble and non-soluble	long-term	Latent period of a few weeks to > 20 years after exposure of a few hours to several years	> 1 year	Chronic beryllium disease: predominantly interstitial granulomas in the lungs, progressive in severity	I	Emphysema; spontaneous pneumothorax
					II	Pulmonary insufficiency
					III	Pulmonary heart[d]; cardio-vascular insufficiency

Table 15 (continued)

Specification of beryllium	Type of exposure	Manifestation of disease	Duration	Clinical form	Degree or stages	Outcome, remote effects and complications
Soluble, poorly soluble and non-soluble	long-term	Latent period of a few weeks to > 20 years after exposure of a few hours to several years	> 1 year	Chronic toxic bronchitis; chronic beryllium disease	I	Emphysema
					II	Chronic bronchitis; bronchiectasia
					III	Pulmonary insufficiency; pulmonary heart; cardio-vascular insufficiency; pneumosclerosis

[a] Adapted from: Burnazian (1983); modified and supplemented.
[b] Brief, but massive exposure.
[c] Less intensive but prolonged exposure resulting also in acute beryllium disease.
[d] Cor pulmonale (right-sided heart failure).

year's natural duration and to exclude those syndromes lasting more than one year".

(a) *Skin effects*

Depending on individual susceptibility, direct contact with soluble beryllium compounds may cause contact dermatitis characterized by reddened, elevated, or fluid-filled lesions on exposed surfaces of the body. This has not been seen in workers handling beryllium hydroxide, pure beryllium, and vacuum-cast beryllium metal (McCord, 1951; NIOSH, 1972). The symptoms develop after a latent period of 1–2 weeks indicating a delayed allergic reaction, and a concomitant conjunctivitis may occur. After cessation of exposure, the skin eruptions usually disappear, whereas, on continued exposure, bronchitis and pneumonitis may develop. Sensitized individuals react much more rapidly and to smaller amounts of beryllium (Van Ordstrand et al., 1945).

A patch test developed by Curtis (1951) appeared to be sensitizing. Eight out of 16 individuals, without previous exposure, developed eczemas from the test itself; also pulmonary exacerbations of beryllium disease were related to this test. Thus, it has not been much used for diagnostic purposes (Curtis, 1959; Reeves, 1986).

When soluble or insoluble beryllium compounds, in crystallized form, are introduced into, or beneath, the skin, e.g., as a result of abrasions or cuts, chronic ulcerations develop, with granulomas appearing, often after several years (Van Ordstrand et al., 1945; Lederer & Savage, 1954). Epstein (1967) classified this reaction as granulomatous hypersensitivity. The granulomas are usually painless. After removal of the beryllium crystals or excision of the granulomatous mass, recovery usually takes place within 2 weeks (NIOSH, 1972).

(b) *Respiratory effects*

Acute respiratory effects produced by beryllium were first reported in beryllium-extraction plants in the Federal Republic of Germany, Italy, the USA, and the USSR. Several cases occurred as a result of the inhalation of soluble beryllium salts, typically the fluoride, at concentrations usually greater than 100 µg Be/m^3.

At a symposium held in 1947 at Saranac Lake, New York (Vorwald, 1950), about 500 cases of acute beryllium disease, with about one dozen deaths, were reported (Eisenbud, 1982). As a consequence, field studies were carried out in the USA, and it soon became evident that the acute respiratory effects could be caused by inhalation of beryllium fluoride, sulfate, chloride, oxide, or hydroxide, and metallic dust (Eisenbud et al., 1948). The physical and chemical properties of the compounds determine the toxicity of the associated beryllium ion. In contrast to low-fired beryllium oxide, no cases of acute beryllium disease were observed in workers exposed to high-fired beryllium oxide (1540 °C).

Eisenbud et al. (1948) reported that all their cases were related to beryllium fluoride and sulfate, the most toxic beryllium compounds, and were associated with concentrations exceeding 100 μg Be/m^3. Concentrations of 1 mg/m^3 consistently produced acute symptoms among almost all exposed workers. A group of 8 workers exposed to beryllium sulfate at concentrations of not more than 15 μg Be/m^3 (analysed spectrographically) did not develop acute disease. These observations served as a basis for a maximum recommended peak value of 25 μg Be/m^3.

Since the adoption of this value by the Atomic Energy Commission and the American Industrial Hygiene Association in the early 1950s (section 5.3.2), cases of acute beryllium disease have dramatically decreased in the USA. The US Beryllium Case Registry (a central file on reported cases of acute and chronic beryllium disease, which was established in 1952) included 224 cases of acute disease, registered up to 1983 (Eisenbud & Lisson, 1983). Most of these cases occurred prior to 1949 and were associated with high mortality. Between 1950 and 1967 only 15 non-fatal cases were reported, all from beryllium production plants. An 18-year-old man developed acute respiratory disease a few days after exposure to the grinding of dies containing a copper-beryllium alloy (Hooper, 1981).

The signs and symptoms of acute beryllium disease range from a mild inflammation of the nasal mucous membranes and pharynx, i.e., rhinitis and pharyngitis, to tracheobronchitis and, depending on the degree, duration, and type of exposure, to severe chemical pneumonitis (NIOSH, 1972; Constantinidis, 1978).

Symptoms of acute pneumonitis, such as progressive cough, shortness of breath, substernal discomfort or pain, appetite and weight loss, general weakness and tiredness, cyanosis, and crepitation, usually occur within 3 days following a massive short-term exposure or within weeks following prolonged exposure to lower concentrations of beryllium.

Chest radiographs show diffused haziness of both lungs, development of soft irregular infiltration areas with prominent peribronchial markings, and the appearance of discrete, large or small nodules, similar to those found in chronic beryllium disease or sarcoidosis (NIOSH, 1972). Pathological studies on tissue samples from 6 patients revealed nongranulomatous acute or subacute pulmonary oedema.

In severe cases, patients died of acute pneumonitis, but in most cases, after cessation of exposure, complete recovery occurred within 1-4 weeks. On re-exposure to beryllium, pneumonitis may appear again (Constantinidis, 1978). In a few cases, chronic beryllium disease developed years after recovery from the acute form (Hardy, 1965).

9.2.1.2 Chronic disease

Chronic beryllium disease differs from the acute form (Table 15) in having a latent period that can vary from several weeks to more than 20 years. It is of long duration, progressive in severity, and with manifestations that have frequently been described as "systemic" (Tepper et al., 1961). However, often, the systemic nature of beryllium disease has been overemphasized, creating the impression that inhalation exposure to beryllium caused whole body poisoning involving all organs of the body. In reality, the manifestations of chronic beryllium disease are entirely consistent with an allergic inflammation of pulmonary tissue in which all effects involving other parts of the body are secondary. Recent evidence (Reeves & Preuss, in press) indicates that chronic beryllium disease may represent a case of "compartmentalized" immune response involving only the alveoli, and resembles other types of hypersensitivity pneumonitis.

Hardy & Tabershaw (1946) were the first to relate the chronic lung disease observed in 17 workers in a fluorescent lamp plant to the inhalation of beryllium. In most of these cases, symptoms, such as dyspnoea on exertion, cough, and weight loss, appeared several months, or even years, after the last exposure.

The disease was first called "delayed chemical pneumonitis" (Hardy & Tabershaw, 1946). After the role of beryllium as causative agent had been confirmed, the term "berylliosis" became widely used. However, this term is considered misleading by some authors (Tepper et al., 1961), first, because it gives the false indication that the beryl ore is involved, and second, because this disease differs from a typical pneumoconiosis owing to its systemic features. The term chronic beryllium disease is therefore preferable.

Of the 888 cases registered in the US Beryllium Case Registry (section 9.2.1.1), 224 cases were classified as acute, 42 cases were unaccounted for, and 622 cases were classified as chronic, 557 of these being due to occupational exposure (Eisenbud & Lisson, 1983). The majority of these were either from exposures within the fluorescent lamp industry (319 cases) or within beryllium extraction plants (101 cases).

As with acute beryllium disease, cases of the chronic form dramatically declined among workers who had started work in the beryllium industry after the implementation of preventive measures, but chronic beryllium disease still occurs. In the Ohio production plant, the incidence rate was reduced from 27 cases per 3000 (1940–60) to 2 per 3000 (1960–83) in newly hired employees. These last 2 cases were attributed to accidentally high exposures (Preuss, 1985).

The United Kingdom Case Registry 1945-85 numbered 49 cases of chronic and 2 cases of acute beryllium disease; 21 beryllium workers had died by 1985, most of them from respiratory failure, 3–29 years after diagnosis (Jones Williams, 1985). In 1988, the United-Kingdom Registry consisted of 60 cases, indicating that new cases were still occurring (Jones Williams, 1988). Four cases of chronic disease developed from acute beryllium disease.

In another British study (Cotes et al., 1983), 8 cases of chronic beryllium disease were recorded in 1963 (6 cases) and 1977 (2 cases). According to air analyses conducted between 1952, the first year of operation, and 1960, beryllium concentrations were thought to be generally far below 2 µg/m^3. However, the occurrence of higher work-place levels cannot be excluded, and this is supported by the observation of 2 cases of acute beryllium pneumonitis.

In Japan, 7 cases of chronic beryllium disease occurred between 1973 and 1975 (Izumi et al., 1976). All were related to exposure to beryllium oxide in a ceramic factory.

Cullen et al. (1987) reported the results of a clinical-epidemiological investigation concerning work-places and employees of a precious metal refinery in Connecticut, USA, engaged in refining and reclaiming beryllium-containing waste materials. In 1983, time-weighted average personal air samples, showed a mean value of 1.2 µg Be/m^3, with a range of 0.22–42.3 µg/m^3. Beryllium concentrations for furnace tenders, sweepers, and dry pan operators were uniformly below 2 µg/m^3, while those for samplers, ball mill operators, and crushers often exceeded this value. Thus, it is surprising that 4 workers, who had worked in the furnace area between 1964 and 1977, developed chronic beryllium disease 4–8 years after the onset of employment. In the higher exposure areas, only one worker developed chronic beryllium disease, diagnosed in 1985. These data suggest that the fume from high temperature operations is more pathogenic than metal dust and that workers who smelt, burn, refine, or weld beryllium or its alloys may still be at risk from chronic beryllium disease, even if exposure concentrations are below the present adopted standards. However, it cannot be completely ruled out that beryllium concentrations were much higher during the period of exposure (1964–77) than those measured in 1983, even though this seems unlikely, because virtually no structural changes or changes in work practices had occurred over the 20-year period.

Infante et al. (1980) also reported a case of chronic beryllium disease diagnosed in an individual who had been exposed to extremely low levels of beryllium (less than 2 µg/m^3) at a rolling mill plant. He was initially exposed in 1965 and was diagnosed in 1972.

Kreiss et al. (1989) conducted a survey on 58 machinists who were exposed to beryllium levels near the current standard. The authors administered a questionnaire, reviewed current medical X-rays, and conducted pulmonary function tests and a peripheral blood lymphocyte transformation test (LTT) on 51 volunteers. Six had abnormal LTT results, and 5 out of 6 sensitized workers agreed to clinical and diagnostic evaluation. Four of the 5 sensitized workers, who were evaluated further, had beryllium disease, defined as granulomata on trans-bronchial lung biopsy, and a 3-fold or higher stimulation index by lung lymphocytes to beryllium salts. The 4 cases of beryllium disease were identified among a group of 20 machinists, first employed 10 or more years prior to the study.

Rossman et al. (1988) evaluated the sensitivity and specificity of the LTT in relation to the diagnosis of chronic beryllium disease. They reported the results of the LTT on cells derived from bronchoalveolar lavage and cells derived from peripheral blood among normal individuals, and individuals who had unequivocal beryllium disease, probable beryllium disease, or sarcoidosis. A stimulation index of more than 5 times control values was considered a positive response with regard to results with cells derived from bronchoalveolar lavage; 14 out of 14 patients with unequivocal beryllium disease had a positive LTT and 3 individuals with probable chronic beryllium disease had a positive LTT. The LTT was negative in 6 beryllium workers who did not have beryllium disease, and also negative in 6 normal volunteers and in 16 patients diagnosed as having sarcoidosis with no history of exposure to beryllium. These results suggest a high degree of sensitivity and specificity for the LTT based on bronchoalveolar lavage.

When cells were derived from peripheral blood, 6 of the 14 patients (42%) with unequivocal chronic beryllium disease had a positive LTT. The peripheral blood LTT was negative for all the remaining patients in the study (personal communication, M.D. Rossman). The LTT results based on peripheral blood cells suggest that only about half of those with chronic beryllium disease may be identified through tests based on lymphocytes derived from blood. These findings taken together with those of Kreiss et al. (1989), who performed LTTs using peripheral blood, also suggest that the chronic beryllium sensitivity in older workers may be more than the 20% observed in the latter study.

In the series of patients studied by Rossman et al. (1988), there was a beryllium worker, in addition to those mentioned above, who had a history of accidental high exposure and showed typical noncaseating lung granulomas on transbronchial biopsy, but no clinically manifest disease, according to physiological and radiological evaluation. His lymphocyte blast transformation was positive. This case may represent a subclinical stage of chronic beryllium disease leading eventually to manifest illness.

The latter finding was confirmed by Newman et al. (1989) who studied 8 workers in an aerospace applications plant and 4 workers in a ceramics manufacturing plant. Radiographic and physiological measurements did not reveal evidence of impairment but showed histopathological pulmonary changes and immunological alterations (bronchoalveolar lavage LTT positive) consistent with chronic beryllium disease. Thus, these cases were considered as "subchronic beryllium disease". In addition, 2 out of 8 beryllium workers in another group with non-beryllium lung disease showed positive LTT and were therefore considered to be "beryllium sensitized".

(a) *Signs and symptoms*

The most common signs and symptoms of the chronic disease are shown in Tables 16 and 17. Pneumonitis associated with dyspnoea on exertion, cough, chest pain, weight loss, fatigue, and general weakness is the most familiar and striking feature (Hardy, 1948; Hardy & Stoeckle, 1959). Right heart enlargement (cor pulmonale) with accompanying cardiac failure, hepatomegaly, splenomegaly, cyanosis, and finger clubbing may also occur (Hall et al., 1959). The appearance of renal stones is quite common, associated with renal colic and dysuria. In spite of the high blood uric acid levels, gout does not occur frequently (Kelley et al., 1969). Stoeckle et al. (1969) found cases of osteosclerosis associated with chronic beryllium disease. Changes in serum proteins and liver function have also been observed.

Andrews et al. (1969) conducted lung function tests on 35 patients. Only 2 cases had normal test results; 11 patients had an interstitial defect, 16 a restrictive defect, and 5 showed evidence of airway obstruction.

Kriebel et al. (1988) studied the sub-clinical effects of beryllium on lung function in beryllium plant workers. After the data were adjusted for age, height, and smoking, decrements in forced vital

Table 16. Signs of chronic beryllium disease[a]

Sign	Frequency (%)
Chest signs	43
Cyanosis	42
Clubbing	31
Hepatomegaly	5
Splenomegaly	3
Complications:	
cardiac failure	17
renal stone	10
pneumothorax	12

[a] Adapted from: Hall et al. (1959).

Table 17. Symptoms of chronic beryllium disease[a]

Symptom	Frequency (%)
Dyspnoea	
on exertion	69
at rest	17
Weight loss	
more than 10%	46
0-10%	15
Cough	
nonproductive	45
productive	33
Fatigue	34
Chest pain	31
Anorexia	26
Weakness	17

[a] Adapted from: Hall et al. (1959).

capacity and forced expiratory volume (in one second) were observed in workers exposed to beryllium for more than 20 years prior to the health survey. These decrements were observed in workers who had no radiographic abnormalities.

Studies of pathological changes in numerous cases of chronic beryllium disease have been reported by Dutra (1948), MacMahon & Olken (1950), Jones Williams (1958), Dudley (1959), Freiman (1959), and Freiman & Hardy (1970). Macroscopically, the lungs may show diffuse changes, affecting all lobes with widespread scattered small nodules and interstitial fibrosis. Epithelioid (sarcoid-like) granulomas are the characteristic feature, together with conspicuous alveolitis. The early granulomatous lesions are aggregates of epithelioid cells surrounded by a poorly defined collection of lymphocytes and plasma cells. Later Langhan-type giant cells develop from the fusion of epithelioid cells. Occasionally, the granulomas fuse to form dense hyalinized nodules.

On light microscopy (Jones Williams, 1958), histochemistry (Williams et al., 1969), and electron-microscopy (Jones Williams et al., 1972), the epithelioid cells, characteristic of the granulomas, were indistinguishable from those in sarcoidosis, Kveim test granulomas, tuberculosis, farmer's lung, and Crohns disease. The same holds true for the appearance of conchoidal (Schaumann) bodies, crystals, and asteroid bodies, which are often numerous in the fibrotic stage of beryllium-induced granulomas.

Dudley (1959) stressed the importance of the diffuse interstitial infiltration of which the granulomas are only a part. It was claimed by Freiman & Hardy (1970) that extensive interstitial change forecast a more severe disease and that it would be a valuable criterion in distinguishing chronic beryllium disease from the very similar granulomatous disease, sarcoidosis.

Skin lesions, resembling those of sarcoidosis, may be present as a secondary response (Tepper et al., 1961). In addition, granulomas may develop in different parts of the body. Högberg & Rajs (1980) found granulomatous myocarditis as the cause of death in two beryllium workers.

Jones Williams & Kilpatrick (1974) reported one case in which local implantation of beryllium led to the generalized disease. A beryllium worker originally had an injury to his hand contaminated with beryllium oxide. A chronic ulcer developed and the affected finger had to be amputated. After several months, ulcerative nodules developed on the forearm and later the lungs were also affected. The patient suffered from dyspnoea. Electron microscopic examination revealed granulomas in the arm and lung lesions, which disappeared after treatment with corticosteroids. However, inhalation exposure to beryllium, due to failures in the exhaust ventilation, cannot be excluded.

Jones Williams et al. (1988) reported skin lesions in 26 beryllium workers in the United Kingdom. Fourteen cases were diagnosed as chronic beryllium disease. Of these, 8 had skin lesions only, and 6 had both skin and lung disease.

The evolution of chronic beryllium disease is not uniform. In some cases, spontaneous alleviation for weeks or years is encountered, followed by exacerbations. In the majority of cases, progressive pulmonary disease occurs with an increased risk of death from cardiac or respiratory failure. The reported morbidity rates among the beryllium workers vary from 0.3 to 7.5%. The length of the latent period and the severity of chronic beryllium disease are in reverse proportion. Bencko & Vasilyeva (1983) reported that latent periods of less than 1 year resulted in fatality rates as high as 37%, while in patients with a latent period of 5–10 years, the mortality rate was only 18%.

(b) *Mechanism of chronic beryllium disease*

The absence of dose-response relationships and the observation that very low inhalation exposure may provoke chronic beryllium disease in sensitized subjects suggest that an immunological mechanism is involved. In 1951, Sterner & Eisenbud (1951) developed a concept for the pathogenesis of beryllium disease based on the hypothesis that "the essential mechanism is a modified immunological reaction". At the same time, Curtis (1951) developed a patch test that gave a positive response in many cases of beryllium disease (Curtis, 1959) and in beryllium-exposed workers (Nishimura, 1966).

Resnick et al. (1970) found an increased concentration of the immunoglobulin fraction IgG in subjects who had had either the cutaneous or the chronic pulmonary forms of beryllium disease. Increased concentrations of IgG, IgA, and IgM were observed by Bencko et al. (1980) in workers in 2 Czechoslovakian power plants (section 9.1), who were exposed to up to 8 μg Be/m^3. However, because of confounding factors, these findings cannot be regarded unequivocally as a specific humoral antiberyllium reaction.

In several studies, an antibody has not been found in the serum of patients with beryllium disease (Voisin et al., 1964; Pugliese et al., 1968; Resnick et al., 1970). However, there is increasing evidence that the delayed cutaneous and granulomatous hypersensitivity are cell-mediated (section 8.7.2). This is supported, not only by the cutaneous sensitivity reaction, but also by lymphoblast transformation, and the production of a migration inhibitory factor by lymphocytes (Reeves & Preuss, 1985). It is also supported by the skin sensitivity in guinea-pigs that is transferred by lymphocytes, but not by serum (Alekseeva, 1966; section 8.7.2).

The guinea-pig could serve as a model to explain the "negative" dose-response relationship in man. Workers with relatively high exposure over several years sometimes developed immunity, whereas with very short work-place exposure and in the "neighbourhood cases", only marginal exposure led to chronic beryllium disease (Reeves & Preuss, 1985).

Reeves & Preuss (1985) suggested that the reactive species is always solid-state and is the oxide with a high density of surface electrostatic charges. Particles of beryllium metal become active through surface oxidation; ionic beryllium, once entered into buffered tissue, precipitates to beryllium hydroxide, which, in turn, ages to form the oxide. Reeves & Preuss (1985) also suggested that it is an adsorptive beryllium-protein complex rather than an ion-bond proteinate that acts as the proximate antigen.

The considerable variability in latency and the lack of dose-response relationships may be explained by immunological sensitization. Acute infection, surgery, pregnancy, or other conditions have often been observed to precede the onset of clinical symptoms of beryllium disease (Hardy, 1965; 1980; Clary et al., 1972). In particular, pregnancy seems to be a precipitating condition; 66% of 95 females registered among the fatal cases in the US Beryllium Case Registry were pregnant (Hardy, 1965).

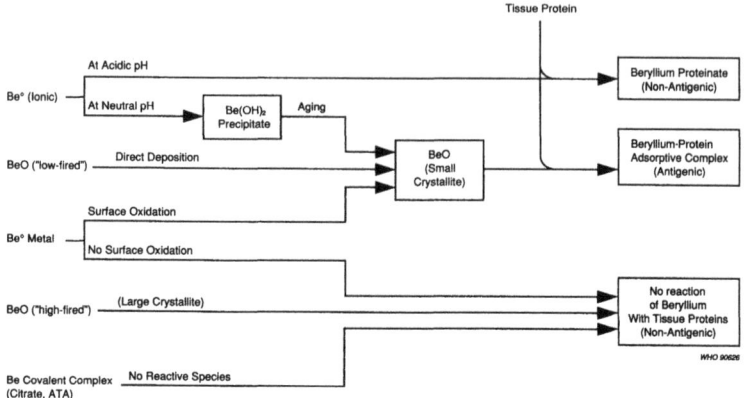

Fig 1. Antigenicity of chemical forms of beryllium
From: Reeves & Preuss (1985)

(c) *Diagnosis*

Drury et al. (1978) and Reeves (1986) have summarized the diagnostic criteria. Since some kind of exposure to beryllium must have preceded the disease, the establishment of exposure by history taking and tissue analysis should serve as a basis for the recognition of chronic beryllium disease, though the presence of beryllium in biological material does not prove disease.

Clinical criteria that indicate beryllium disease include scattered opacities on chest X rays, impaired lung function, interstitial pneumonitis, and systemic toxicity. However, other types of interstitial lung disease show similar pathophysiological and radiological features. Differentiation between chronic beryllium disease and sarcoidosis is most difficult.

The patch test developed by Curtis (1951) is not recommended for the diagnosis of beryllium disease, because it can give false positives and negatives. Moreover, it may itself induce a skin sensitivity reaction or flare-ups in dormant pulmonary lesions.

The macrophage migration inhibition assay and the lymphoblast transformation test are useful indicators of beryllium hypersensitivity (Reeves & Preuss, 1985). Both *in vitro* tests are gaining importance as useful methods for the diagnosis of beryllium disease. Using the lymphocyte transformation test, Deodhar et al. (1973) and Preuss et al. (1980) reported positive results in about 70% of beryllium disease patients. Jones Williams & Williams (1983) found a 100% positive response in 16 patients with chronic beryllium disease, and a negative response in 10 subjects who were suspected of having chronic beryllium disease. Only 2 out of 117 healthy beryllium workers had a positive response.

The lymphocyte transformation test was also 100% positive, independent of steroid therapy, and reproducible in 7 patients (Williams & Jones Williams, 1982). In contrast, the macrophage migration inhibition test was only positive in 4 patients (57%) who were not on steroids, and was not reproducible.

Bargon et al. (1986) suggested that lymphocyte transformation test results should only be considered as evidence of beryllium disease if there were positive test results over a wide range of concentrations. Of 23 foundry workers, 3 showed clear positive results. Two of these workers were diagnosed as having diffuse granulomatous lung disease; the other worker did not show any signs of an interstitial lung disease. In the other exposed workers and 20 non-exposed controls, lymphocyte transformation test results were either negative or only positive at one or two concentrations.

Rossman et al. (1988) have reported the development of a fairly reliable test for chronic beryllium disease. The test measures the proliferative response of bronchoalveolar lymphocytes to beryllium. In 14 patients diagnosed with chronic beryllium disease, the bronchoalveolar lymphocytes showed a positive dose-related proliferative response to beryllium sulfate. This response was not observed in 6 normal volunteers, 16 patients with sarcoidosis, or 4 beryllium workers proved not to have chronic beryllium disease.

Rossman et al. (1988), Newman et al. (1989), and Saltini et al. (1989) confirmed that the lymphocyte transformation test is generally more sensitive and specific using lymphocytes from bronchoalveolar lavage than using peripheral blood lymphocytes.

Based on the findings of beryllium-related pathological and immunological alterations in the absence of radiographic or physiological impairment, Newman et al. (1989) suggested the following classification:

- Beryllium-sensitization: persons with a positive blood and/or lung lymphocyte transformation test;
- Subclinical beryllium disease: persons with additional pathological alterations on biopsy, but who are asymptomatic;
- Clinical beryllium disease: persons who meet all the diagnostic criteria and have clinical symptoms or measurable impairment.

The lymphocyte transformation test, especially if performed on bronchoalveolar lavage, appears to be the method of choice for the detection of beryllium hypersensitivity. Workers with positive results should be considered for removal from further exposure (see section 11).

9.3 Carcinogenicity

9.3.1 Epidemiological studies

Hardy et al. (1967) reviewed the mortality and morbidity data of the US Beryllium Case Registry for the period 1952–66. They did not find any evidence that beryllium caused cancer in human beings. Likewise, Stoeckle et al. (1969) reviewing the clinical findings and course in 60 patients with chronic beryllium disease noted 17 deaths, but none due to cancer. Mancuso & El-Attar (1969) introduced an epidemiological approach in the study of the mortality pattern of beryllium workers at 2 separate companies. The study was based on the social security files of 3685 white males employed from 1937 to 1948. From 729 deaths up to the year 1966, 31 were due to lung cancer. Because of several principal limitations, particularly the small numbers in the 160 subcategories into which the 729 deaths were distributed, it was not possible to make any trend statements or statistical tests, leaving open the question of carcinogenic risk.

In the study by Mancuso (1970), a subgroup of individuals at the Ohio plant, who had clinical case histories of beryllium-induced acute bronchitis or pneumonitis, had been identified between

1940–48 and were followed-up until 1967. A higher rate of lung cancer was noted among the 145 workers previously diagnosed with acute bronchitis or pneumonitis compared with the rate among those without acute illness. Six out of 8 (75%) of the lung cancer cases identified in the Ohio cohort up to 1967 came from the group of 145 workers diagnosed with acute pneumonitis, which represented less than 15% of the cohort.

Bayliss et al. (1971) studied workers in the beryllium-processing industry in Ohio and Pennsylvania. They observed a slightly elevated risk of lung cancer, though the results were not statistically significant.

Between 1979 and 1980, reports of these cohorts were updated by Mancuso (1979; 1980) and by Wagoner et al. (1980). A formal epidemiological study of individuals entered in the US Beryllium Case Registry (BCR) was reported by Infante et al. (1980). These investigations serve as the basis for the analysis of the mortality patterns of workers exposed to beryllium.

Criticisms have been raised (US EPA, 1987) about the interpretation of the excess lung cancer risk observed in the studies by Mancuso (1979; 1980), Wagoner et al. (1980), and Infante et al. (1980). The major concerns relate to selection bias, confounding from cigarette smoking, and underestimation of the expected number of lung cancer deaths reported in the original studies. These issues will be dealt with in the following review. In particular, it should be noted that the data presented here are based on the ratios of observed versus expected deaths, as recalculated by Saracci (1985), taking into consideration the increased incidence of lung cancer between 1968 and 1975 in the total US population.

Mancuso (1979) studied all white males employed at some time between 1 January 1942 and 31 December 1948 at the Ohio and Pennsylvania facilities. Cohort members were identified by payroll records submitted quarterly by the employer to the US Social Security Administration. These cohorts were followed up to 1974 for Ohio and up to 1975 for Pennsylvania. Observed deaths from lung cancer were compared with the number of deaths using United States national mortality rates adjusted for age and calendar time periods. However, rates for 1965–67 were also used to estimate expected mortality for the years 1968–75. This procedure resulted in an underestimation of expected lung cancer mortality by about 10% (Saracci, 1985). Thus, the expected mortality from lung cancer reported in Mancuso (1979) has been increased by 10%. The observed (O) deaths from lung cancer together with the "adjusted" expected (E) number of lung cancer deaths for the Ohio and Pennsylvania facilities are shown in Table 18. Among the Ohio workers, the overall risk ratio of lung cancer was 1.8 (95% confidence interval (CI) is 1.2–2.7). (When the lower number of the CI is 1.0 or more, the risk ratio is significantly elevated). The significantly elevated overall risk of lung cancer is confined to individuals who were employed for less than one year and 1–4 years and were followed for more than 15 years from initial employment. For those followed for more than 15 years, the risk ratio is 2.0 (95% CI = 1.3–3.1).

As shown in Table 18, the cohort of Pennsylvania employees had an overall risk of 1.2 (95% CI = 0.9-1.7). However, for individuals followed for more than 15 years, the risk ratio was 1.5 (95% CI = 1.0–2.1). Consistent with the observations from the Ohio employees, the Pennsylvania beryllium workers' elevated risk of lung cancer was confined to individuals, employed for less than one year or for 1–4 years, who had been followed for 15 or more years since initial employment.

Mancuso (1980) followed both the Ohio and Pennsylvania cohorts up to 1976 and pooled the data from both cohorts. He compared their expected lung cancer mortality with that expected on the basis of the mortality rates of other industrial workers who were employed in the same geographical area and for similar periods of time and were followed for a similar calendar time period. As shown in Table 19, the risk ratios for lung cancer were significantly elevated whether the expected was based on "all viscose rayon" employees or on those who had never transferred from their department of initial employment. The lung cancer risk ratios range from 1.4 to 1.6 and are statistically significant. If the data shown in Table 18 are combined, the overall risk of lung cancer is 1.4 with expected mortality based on the US general population rates. Thus, the risk range of 1.4–1.6 with expected mortality based on the viscose rayon employees is similar. Mancuso (1980) did not present data on cigarette smoking among his beryllium cohort members. However, an indirect indication that cigarette smoking may not have played

Table 18. Observed (O) and expected (E) deaths due to lung cancer and their ratios with 95% confidence interval (CI) according to duration of employment and time since onset of employment in two US beryllium production facilities at some time between 1 January 1942 and 31 December 1948, followed through 31 December 1974[a]

Interval since onset of employment (years)	< 1			1-4			> 5			Total		
	O/E	Ratio	CI	O/E	Ratio	CI	O/E	Ratio	CI	O/E	Ratio	CI
(i) *Ohio facility*												
< 15	3/1.96	1.5	0.3-4.5	0/0.70	0	0-5.3	0/0.26	0	0-14.2	3/2.92	1.0	0.2-3.0
≥ 15	14/7.14	2.0	1.1-3.3	5/1.91	2.6	0.8-6.1	3/1.79	1.7	0.3-4.9	22/10.84	2.0	1.3-3.1
Total	17/9.10	1.9	1.1-3.0	5/2.61	1.9	0.6-4.4	3/2.05	1.5	0.3-4.3	25/13.76	1.8	1.2-2.7
(ii) *Pennsylvania facility*												
< 15	3/4.70	0.6	0.1-1.9	1/2.11	0.5	0.1-2.6	0/0.98	0	0-4.1	4/7.79	0.5	0.1-1.3
≥ 15	23/14.12	1.6	1.0-2.4	10.5/5.80	1.7	0.8-3.2	3/4.30	0.7	0.1-2.0	36/24.22	1.5	1.0-2.1
Total	26/18.82	1.4	0.9-2.0	11/7.91	1.4	0.7-2.5	3/5.28	0.6	0.1-1.7	40/32.01	1.2	0.9-1.7

[a] Adapted from: Saracci (1985); original study from Mancuso (1979).

Table 19. Observed (O) deaths due to lung cancer among 35 to 74-year-old workers of two US beryllium production facilities as contrasted with those expected (E) on the basis of two cohorts of workers in the viscose rayon industry, employed for similar durations of time and followed over the same period of time[a]

Duration of employment (months)	Lung cancer mortality[b]					
	O/E[c]	Ratio	CI	O/E[d]	Ratio	CI
≤12	52/37.60	1.4	0.9-2.1	52/31.67	1.6	1.1-2.6
13-48	14/13.26	1.1	0.5-2.2	14/10.82	1.3	0.6-2.9
49	14/ 6.32	2.2	0.9-5.7	14/ 8.14	1.7	0.7-4.1
Total	80/57.18	1.4	1.0-2.0	80/50.63	1.6	1.1-2.2

[a] Adapted from: Saracci (1985); original study from Mancuso (1980).
[b] CI = 95% confidence interval.
[c] All viscose rayon employees.
[d] Viscose rayon employees who had never transferred from department of initial employment.

a major role in the increased lung cancer risk among beryllium employees can be derived from the observation that, when a population of industrial workers was used to compute expected mortality, the elevated lung cancer risk remained about the same. It could be presumed that blue collar workers, employed in the same geographical area over the same calendar time period, who had similar length of employment patterns, would have similar smoking habits. Thus, it is considered unlikely that cigarette smoking played a major role in the excess lung cancer risk.

Wagoner et al. (1980) conducted a cohort study on beryllium production workers employed at the Pennsylvania facility that was studied by Mancuso (1979, 1980). However, Wagoner et al. (1980) identified their cohort members through company records and a medical survey conducted at the facility in 1968. The cohort consisted of 3055 white males who were employed for some time during the period from January 1942 to September 1968. The cohort was followed through 1975. Observed mortality was compared to the expected, based on the US white male general population, adjusted for age and calendar time period. However, lung cancer rates for 1965–67 were also used to estimate expected mortality for the years 1968–75. Therefore, the expected lung cancer mortality, as reported in the Wagoner et al. (1980) paper, has been increased by 10%. The results for lung cancer are shown in Table 20. For the total group, 47 lung cancer deaths were observed compared with 37.72 expected. The risk ratio was 1.2 (95% CI = 0.9-1.7). For those followed for 25 or more years since initial employment, the ratio was significantly elevated, O/E = 1.7 (95% CI = 1.0-2.6). The risk of lung cancer also increased with an increase in latency, but not with duration of employment. NIOSH re-analysed the data using updated lung cancer rates (Foege, Personal communication, 1981). The results demonstrated a significant increase in lung cancer deaths for those with more than 15 years of latency (39 observed versus 13.36 expected, $P = 0.035$).

Table 20 provides data for mortality from non-malignant lung disease (excluding influenza and pneumonia). Overall, the risk ratio was 1.6 (95% CI = 1.1–2.3). The excess appears to be restricted to those employed for less than 5 years, O/E = 1.9 (95% CI = 1.2–2.7). A significant excess of mortality was seen for heart diseases. There were 396 deaths observed compared with 349.32 expected. The risk

Table 20. Observed (O) and expected (E) deaths and their ratios with 95% confidence interval (CI) due to (i) lung cancer and (ii) non-neoplastic respiratory disease, according to duration of employment and time since onset of employment in a US beryllium production facility (Pennsylvania) at some time between January 1942 and September 1968, followed through 1975[a]

Interval since onset of employment (years)	Duration of employment (years)[b]								
	<1			14			Total		
	O/E	Ratio	CI	O/E	Ratio	CI	O/E	Ratio	CI
(i) Deaths due to lung cancer									
<15	8/8.74	0.9	0.4-1.8	1/1.63	0.6	0.1-3.8	9/10.37	0.9	0.4-1.6
15-24	16/12.72	1.2	0.7-1.9	3/2.76	1.1	0.2-3.5	18/15.48	1.2	0.7-1.8
≥25	17/9.98	1.7	1.0-2.7	3/1.89	1.6	0.4-5.1	20/11.87	1.7	1.0-2.6
Total	41/31.44	1.3	0.9-1.7	7/6.28	1.1	0.4-2.3	47/37.72	1.2	0.9-1.7
(ii) Deaths due to non-neoplastic respiratory disease									
<15	6/3.57	1.7	0.6-3.7	1/0.66	1.5	0.1-8.4	7/4.23	1.7	0.7-3.4
15-24	11/6.36	1.7	0.9-3.1	1/1.41	0.7	0.1-4.0	12/7.77	1.6	0.8-2.7
≥25	12/5.62	2.1	1.1-3.7	0/1.15	0	0-3.2	12/6.77	1.8	0.9-3.1
Total	29/15.55	1.9	1.2-2.7	2/3.22	0.6	0.1-2.2	31/18.77	1.6	1.1-2.3

[a] Adapted from: Saracci (1985), original study from Wagoner et al. (1980).
[b] Employment histories ascertained only for 1967-68.

ratio was 1.1 (95% CI = 1.0-1.2). A large number of these individuals died from right-sided heart failure (cor pulmonale) as a consequence of beryllium lung disease. These data are not shown in tabular form.

Concern has arisen that this study may have bias in the selection of cohort members for study. However, the results for lung cancer in the Wagoner et al. (1980) study of the Pennsylvania cohort, as shown in Table 20, are virtually the same as those reported in Mancuso's study of the same facility, using records on employment derived from the social security administration.

With regard to the effect of cigarette smoking on the increased lung cancer risk, it has been estimated that difference in smoking habits between the cohort members and the general population may account for about 4% of the increased risk (Saracci, 1985).

Infante et al. (1980) studied a group of 421 white males who had been entered with the US Beryllium Case Registry, while alive, between 1 July 1952 and 31 December 1975. The cohort was followed through 1975. The NIOSH modified life-table method was used to calculate expected mortality. Again, because of the problem of underestimation of expected lung cancer mortality, the expected numbers of deaths have been increased by 10%. The results for lung cancer and for death from non-malignant respiratory disease are shown in Table 21. A significantly elevated risk ratio for lung cancer (6 observed versus 2.10 expected, 95% CI = 1.0–6.2) was observed among those who were entered in the Registry with a diagnosis of beryllium-induced acute pneumonitis or bronchitis. No excess of lung cancer was observed among those entered with chronic respiratory disease (1 observed death versus 1.52 expected, 95% CI = 0.7–3.7). However, the overwhelming excess of mortality due to non-neoplastic respiratory disease (42 observed deaths, 0.65 expected, 95% CI = 46.6-87.3) may have limited the ability to detect an excess of lung cancer in this latter group. Another interesting observation is that those who were entered in the Registry with a diagnosis of acute respiratory illness had a 10-fold increased risk of dying from chronic non-neoplastic respiratory disease. The risk ratio is 10.3 (95% CI = 4.9-18.9) as shown in Table 21.

Table 21. Observed (O) and expected (E) deaths and their ratios with 95% confidence interval (CI) due to lung cancer and non-neoplastic respiratory disease among white males enrolled in the US Beryllium Case Registry, while alive, between 1 July 1952 and 31 December 1975[a]

Interval since initial beryllium exposure (years)	Lung cancer			Non-neoplastic respiratory disease[b]		
	O/E	Ratio	CI	O/E	Ratio	CI
(i) Acute respiratory illness group (N = 223)						
<15	1/0.38	2.6	0.1-14.7	1/0.14	7.1	0.2-39.8
≥15	5/1.72	2.9	0.9-6.8	9/0.83	10.8	5.0-20.6
Total	6/2.10	2.9	1.0-6.2	10/0.97	10.3	4.9-18.9
(ii) Chronic respiratory disease group (N = 198)						
<15	0/0.155	0	0-24.6	9/0.05	180.0	82.3-341.7
≥15	1/1.36	0.7	0.1-4.1	33/0.60	55.0	37.9-77.2
Total	1/1.52	0.7	0.1-3.7	42/0.65	64.4	46.6-87.3

[a] Adapted from: Saracci (1985); original study from Infante et al. (1980).
[b] Excludes influenza and pneumonia.

10. EVALUATION OF HUMAN HEALTH RISKS AND EFFECTS ON THE ENVIRONMENT

10.1 Evaluation of human health risks

Beryllium is widely distributed in the environment, generally occurring in trace quantities. The growing use of this element in high technology applications increases the potential for exposure to beryllium in its various forms, particularly beryllium metal, beryllium oxide, and beryllium-containing alloys. Inhalation exposure is the most significant route in terms of risk of adverse health effects. Skin contact with beryllium metal and its compounds is also of concern.

Provided that control measures in the beryllium industry are adequate, general population exposure today is mainly confined to low levels of airborne beryllium from the combustion of fossil fuels, especially coal. In more exceptional cases, where the beryllium content of the coal being burned is unusually high and no adequate control measures are applied, this source could pose health problems. Tobacco smoking probably contributes to inhalation exposure but, at present, there are only limited data. The use of beryllium for dental protheses must also be taken into consideration, in view of its high sensitization potential. There is a small intake from water and food; from a toxicological point of view, the ingestion of beryllium is of minor importance. However, the present data base is insufficient for a quantitative assessment of beryllium intake via air, food, drinking-water, and tobacco smoke.

Depending on individual susceptibility, direct contact with soluble beryllium salts can cause delayed (contact) dermatitis, occasionally associated with conjunctivitis. When beryllium compounds are retained in, or beneath, the skin, chronic granulomatous ulcerations develop.

Acute effects on the respiratory tract including nasopharyngitis, bronchitis, and severe chemical pneumonitis have been reported as an occupational disease among beryllium workers exposed to high concentrations of fumes or dust, usually exceeding 100 µg

Be/m^3. In particular, beryllium fluoride and sulfate, but also the low-fired oxide, have produced acute poisoning, while the less soluble high-fired oxide has not caused acute beryllium disease. Usually, complete recovery has occurred after removal from the exposure, but in severe cases, patients have died of pneumonitis. After the implementation of preventive measures, cases of acute beryllium disease have drastically decreased and, today, may only occur as a consequence of failures in control systems.

Inhalation of beryllium can also produce chronic beryllium disease, either years after recovery from the acute form or, more commonly, independently, after a latent period varying from several weeks to more than 20 years and, frequently, several years after termination of exposure. The clinical, radiological, functional, and pathological features of chronic beryllium disease resemble those of sarcoidosis, though the interstitial inflammatory reaction tends to be more prominent in chronic beryllium disease. Primarily, the lung is affected. Pulmonary disease associated with dyspnoea on exertion, cough, chest pain, weight loss, and general weakness is the most familiar and striking feature. Effects on other organs may be secondary rather than systemic effects. Epithelioid granulomas with varying amounts of interstitial inflammation form the characteristic microscopic picture. The highest morbidity rates have been found in patients who developed the disease after a latency period of less than one year. No correlation has been found between the intensity of exposure and the severity of the disease. The great variability in latency and the lack of dose-response relationships in chronic beryllium disease may be explained by immunological sensitization. Pregnancy seems to be a precipitating "stress factor". The adoption of exposure standards has clearly decreased the incidence of chronic beryllium disease. However, this disease may still occur among sensitized individuals who have been exposed to concentrations of around 2 μg/m^3.

The available data from genotoxicity tests indicate that beryllium interacts with DNA and causes gene mutations, chromosomal aberrations, and sister chromatid exchange in cultured mammalian somatic cells, though it has been found not to be mutagenic in bacterial test systems.

Intravenous and intramedullary injection of beryllium metal and various compounds produced bone cancer in rabbits, but not in guinea-pigs, rats, and mice. Inhalation or intratracheal exposure to soluble and insoluble beryllium compounds, beryllium metal, various beryllium alloys, or beryl induced lung tumours in rats. No pulmonary tumours have been observed in rabbits, hamsters, or guinea-pigs. On the whole, the carcinogenic activity of beryllium in different animals has been confirmed, though study design and laboratory practice at the time these studies were conducted were mostly not in compliance with the current approaches used in carcinogenicity tests. In particular, the reported exposure data should be considered with caution.

Several epidemiological studies have provided data indicating an excess lung cancer incidence in populations occupationally exposed to beryllium. These data were derived from studies of two United States working populations and a registry of clinical cases partially covering these same populations as well as other occupations. The question pending is whether chance, bias, or confounding, rather than exposure to beryllium, can explain the association (Saracci, 1985).

The cohorts of beryllium workers employed at both the Ohio and Pennsylvania production facilities indicate a statistically significant excess risk of lung cancer after long intervals from initial exposure (15-25 years) among workers with less than 5 years' duration of employment. The data base from one of the studies of the Pennsylvania cohort has been scrutinized closely. The fact that two different approaches to cohort selection (use of social security data by Mancuso (1979) and the use of company records by Wagoner et al. (1980)) result in essentially concordant findings is evidence against the role of major bias in subject selection or response assessment. Another fact against selection bias is that the elevated lung cancer risk ratios were similar in the Mancuso (1980) study when an industrial reference population, selected in a manner similar to the beryllium cohort, was used for comparison rather than the United States white male population (Saracci, 1985).

After chance and bias, confounding from cigarette smoking needs to be considered. The distribution of cigarette smoking was determined from a morbidity survey conducted at the Pennsylvania facility in 1968 and was reported by Wagoner et al. (1980). On the basis of these data, Saracci (1985) estimated that the difference in smoking habits between the beryllium workers and the general population was such as to increase the risk of lung cancer by 4% over the risk in the general population. Other evidence against smoking playing a major role in the elevated lung cancer risk was provided indirectly by Mancuso (1980), who observed a significant increase in lung cancer in the beryllium workers compared with an industrial population from the same geographical area. (Smoking habits and other socioeconomic factors are assumed to be similar in these two groups of blue collar workers). Thus, the observed increase in lung cancer could hardly be accounted for by smoking alone.

If the increased risk of lung cancer is partially or totally related to beryllium exposure, it would be expected that the greater the exposure, the higher the risk. Subjects entered in the Beryllium Case Registry with a diagnosis of beryllium-induced acute pneumonitis or bronchitis had often experienced high exposures. In fact, the Registry data indicate a 3-fold increased risk of lung cancer for these subjects (Infante et al., 1980). Mancuso (1970) also reported an elevated frequency of lung cancer in a subgroup of the Ohio cohort that was identified as having acute beryllium disease.

In the cohort studies on production workers, the increased lung cancer risk appeared to be localized among workers with less than 5 years or less than one year of employment. As exposure in the past, particularly before 1950, was substantially higher, length of employment would be a distorted indicator of the actual exposure accumulated by a worker. This explanation for the elevated risk in these workers appears to be more likely than an alternative one that workers with relatively short employment are a selected group who experience higher mortality from lung cancer. This explanation is further supported by the observation of an increased risk of lung cancer among beryllium workers compared with other industrial workers employed for similar periods of time.

10.2 Evaluation of effects on the environment

Data on the fate of beryllium in the environment are limited. Atmospheric beryllium oxide particles (combustion processes) return to earth by wet and dry deposition. Within the environmental pH range, beryllium is absorbed by finely-dispersed sedimentary minerals preventing release to ground water. Therefore, beryllium concentrations in surface waters (μg/litre range) and soils (mg/kg dry weight range) are usually low and probably do not affect the environment.

Little is known about the effects of beryllium on microorganisms. At high pH, beryllium salts have growth-stimulating effects on algae and on crop plants. There is evidence that beryllium is able to substitute for magnesium in the growth process of crop plants, thus reducing their magnesium requirement. At or below pH 7, beryllium is toxic for aquatic and terrestrial plants, because of its inhibitory effects on enzyme activity and the uptake of essential mineral ions. Most plants take up beryllium in small amounts, but very little is translocated within the plant.

The toxicity of beryllium for aquatic animals increases with decreasing water hardness. In acute toxicity studies on different freshwater fish species, LC_{50} values ranged between 0.15 and 32 mg Be/litre.

No data are available on the effects of beryllium on domestic or wild terrestrial animals.

There is no evidence that beryllium biomagnifies within food chains.

10.3 Conclusions

The health hazards of beryllium are almost exclusively confined to inhalation exposure and skin contact. Except for the accidental release of beryllium into the environment, the general population is only exposed to very low levels of airborne beryllium, which do not pose a health hazard. Because of the high sensitization and allergenic potential of ionic beryllium, the use of beryllium for dental protheses should be reconsidered.

10.3.1. Acute beryllium disease

Occupational exposure to beryllium poses a health hazard that may result in skin lesions and adverse effects on the respiratory tract. Of the latter, acute beryllium disease can be encountered after exposure to relatively high concentrations of beryllium in fumes and dust (> 100 µg/m^3). Because of improved control measures, such high concentrations are not expected to occur in today's occupational settings.

10.3.2 Chronic beryllium disease

Hundreds of cases of chronic beryllium disease have been diagnosed in various countries throughout the world. The vast majority of these cases have been the result of previous exposure to high concentrations of beryllium in the extraction and smelting of beryllium, fluorescent tube production (no longer a source of beryllium exposure), and in beryllium metal production.

More recently, cases of beryllium disease have been diagnosed following low-level exposure (around 2 µg/m^3). The results of recent studies suggest that some degree of immunological responsiveness to beryllium may be common among workers exposed for more than 10 years. Thus, current occupational exposure standards may not exclude the development of chronic beryllium disease in beryllium-sensitized individuals.

10.3.3 Cancer

Multiple studies on experimental animals have provided sufficient evidence that beryllium is carcinogenic. The available epidemiological data lead to the conclusion that beryllium is the most likely single explanation for the excess lung cancer observed in exposed workers.[a]

---

[a] Professor A.L. Reeves dissented from this statement.

11. RECOMMENDATIONS

1. There is a need for well conducted inhalation toxicity studies on experimental animals, focused on the species- and compound-specifity of beryllium carcinogenicity.

2. Mechanistic studies on the immunotoxicity of beryllium are needed.

3. Mechanisms of carcinogenesis should be investigated, including the molecular mechanisms of beryllium transport and binding in cells and cell nuclei.

4. Improvement of analytical methods and the application of quality control are necessary.

5. Reliable data on the beryllium contents of food, drinking-water, and tobacco originating from different parts of the world, are required.

6. In the work-place, regular monitoring of air concentrations of beryllium should be performed.

7. The use of the lymphocyte transformation test (LTT) should be considered for identifying sensitized individuals. These should be permanently removed from further exposure to beryllium.

8. Human data on bioavailability, tissue levels, and body burden are required.

9. Selected human subpopulations should be monitored to determine beryllium exposure and body burden.

10. The contribution of beryllium released from solid rocket propellants and in space technology should be established.

11. Individuals suspected of having sarcoidosis, employed in any occupation, should be evaluated for immunological sensitivity to beryllium, because of the possible unknown exposure to beryllium.

12. The use of beryllium for dental protheses should be reconsidered, because of the high sensitization and allergenic potential of ionic beryllium.

12. PREVIOUS EVALUATIONS BY INTERNATIONAL BODIES

An International Agency for Research on Cancer Working Group (IARC, 1987) evaluated the carcinogenicity of beryllium and assigned beryllium and beryllium compounds to Group 2A, concluding that they are probably carcinogenic to human beings. The evaluation was reported as follows:

"A. Evidence for carcinogenicity to humans (limited)

"Observations, reviewed elsewhere on beryllium-exposed subjects cover two industrial populations and a registry of berylliosis cases. Workers at beryllium extraction, production and fabrication facilities in the USA were followed up and their causes of mortality compared with those of both the general population and a cohort of viscose-rayon workers. Ratios of observed to expected deaths for lung cancer in the two industrial populations (65 observed) were found to be elevated in both comparisons (1.4 in respect of both the general population [95% confidence interval (CI), 1.1–1.8] and the viscose-rayon workers [1.0–2.0]) and tended to be concentrated in workers who had been employed for less than five years. Data from the US Beryllium Case Registry, in which cases of beryllium-related lung diseases were collected from a wide variety of sources (including the two facilities previously mentioned), indicate an approximately three-fold (six deaths observed, 2.1 expected; ratio of observed: expected, 2.9 [95% CI, 1.0–6.2]) increase in mortality from lung cancer among subjects who had suffered from acute berylliosis, which usually follows heavy exposure to beryllium, but not among those who had had chronic berylliosis (one death observed, 1.4 expected; ratio of observed:expected, 0.7; 95% CI, 0.1–3.7).

B. Evidence for carcinogenicity to animals (sufficient)

Beryllium metal, beryllium-aluminium alloy, beryl, ore, beryllium chloride, beryllium fluoride, beryllium hydroxide, beryllium sulphate (and its tetrahydrate) and beryllium oxide all produced lung tumours in rats exposed by inhalation or intratracheally. Single intratracheal instillations or one-hour inhalation exposures were effective. Beryllium oxide and beryllium sulphate produced lung tumours in monkeys after intrabronchial implantation or inhalation. Beryllium metal, beryllium carbonate, beryllium oxide, beryllium phosphate, beryllium silicate and zinc beryllium silicate all produced osteosarcomas in rabbits following their intravenous and/or intramedullary administration.

C. **Other relevant data**

No data were available on the genetic and related effects of beryllium and beryllium compounds in humans.

All of the available experimental studies considered by the Working Group were carried out with water-soluble beryllium salts. In one study, beryllium sulphate increased the frequency of chromosomal aberrations and sister chromatid exchanges in human lymphocytes and in Syrian hamster cells *in vitro;* in another study, chromosomal aberrations were not seen in human lymphocytes. It caused transformation of cultured rodent cells in several test systems. In one study, beryllium chloride induced mutation in cultured Chinese hamster cells. Beryllium sulphate did not induce unscheduled DNA synthesis in rat hepatocytes *in vitro*, mitotic recombination in yeast or mutation in bacteria. Beryllium chloride was mutagenic to bacteria."

REFERENCES

ACGIH (1988) Threshold limit values and biological exposure indices for 1988-1989. Cincinnati, Ohio, American Conference of Governmental Industrial Hygienists Inc., p.12.

ALDRIDGE, W.N. (1950) Beryllium and alkaline phosphatase. Nature (Lond.), 165: 772.

ALDRIDGE, W.N., BARNES, J.M., & DENZ, F.A. (1949) Experimental beryllium poisoning. Brit. J. exp. Pathol., 30: 375-388.

ALEKSEEVA, O.G. (1966) Ability of beryllium compounds to cause allergy of the delayed type. Fed. Proc. (Trans. Suppl.), 25: 843-846.

ANDRE, S., METIVIER, H., LANTENOIS, G., BOYER, M., NOBILE, D., & MASSE, R. (1987) Beryllium metal solubility in the lung, comparison of metal and hot-pressed forms by *in vivo* and *in vitro* dissolution bioassays. Human Toxicol., 6: 233-240.

ANDREWS, J.L., KAZEMI, H., & HARDY, H.L. (1969) Patterns of lung dysfunction in chronic beryllium disease. Am. Rev. respir. Dis., 100: 791-800.

APHA (1971) Standard methods for the examination of water and wastewater. 13th ed. Washington, DC, American Public Health Association, pp. 66-67.

ASAMI, T. & FUKAZAWA, F. (1985) Beryllium contents of uncontaminated soil and sediments in Japan. Soil Sci. plant Nutr., 31: 43-54.

ATSDR (1987) Toxicological profile for beryllium. Atlanta, Georgia, Agency for Toxic Substances and Disease Registry (ATSDR), US Public Health Service, Department of Health & Human Services, 96 pp.

AWADALLAH, R.M., SHERIF, M.K., AMRALLAH, A.H., & GRASS, F. (1986) Determination of trace elements of some Egyptian crops by instrumental neutron activation, inductively coupled plasma-atomic emission spectrometric and flameless atomic absorption spectrophotometric analysis. J. radioanal. nucl. Chem., 98(2): 235-246.

BALLANCE, J., STONEHOUSE, A.J., SWEENEY, R., & WALSH, K. (1978) Beryllium and beryllium alloys. In: Kirk, R.E. & Othmer, D.F., ed. Encyclopedia of chemical technology, 3rd ed., New York, John Wiley & Sons, Vol. 3, pp. 803-823.

BARGON, J., KRONENBERGER, H., BERGMANN, L., BUHL, R., MEIER-SYDOW, J., & MITROU, P. (1986) Lymphocyte transformation test in a group of foundry workers exposed to beryllium and non-exposed controls. Eur. J. respir. Dis., 69: 211-215.

BARNA, B.P., DEODHAR, S.D., CHIANG, T., GAUTAM, S., & EDINGER, M. (1984) Experimental beryllium-induced lung disease. I. Differences in immunologic responses to beryllium compounds in strains 2 and 13 guinea pigs. Int. Arch. Allergy appl. Immunol., 73(1): 42-48.

BARNES, J.M., DENZ, F.A., & SISSONS, H.A. (1950) Beryllium bone sarcomata in rabbits. Br. J. Cancer, 4: 212-222.

BAUMGARDT, B., JACKWERTH, E., OTTO, H., & TOELG, G. (1986) Trace analysis to determine heavy metal load in lung tissue. Int. Arch. occup. environ. Health, 58: 27-34.

BAYLISS, D.L., LAINHART, W.S., CRALLY, L.J., LIGO, R., AYER, H., & HUNTER, F. (1971) Mortality patterns in a group of former beryllium workers. In: Transactions of the 33rd Annual Meeting of the American Conference of Governmental Industrial Hygienists, Toronto, Canada, 24-28 May 1971, pp. 94-107.

BELMAN, S. (1969) Beryllium binding of epidermal constituents. J. occup. Med., 11: 175-183.

BENCKO, V. (in press) Health risk of human exposure to beryllium and nickel. Int. J. Biosoc. Med. Res., 11:

BENCKO, V. & VASILYEVA, E.V. (1983) Hygienic and toxicological aspects of occupational and environmental exposure to beryllium. J. Hyg. Epidemiol. Microbiol. Immunol., 27: 403-417.

BENCKO, V., BREZINA, M., BENES, B., & CIKRT, M. (1979a) Penetration of beryllium through the placenta and its distribution in the mouse. J. Hyg. Epidemiol. Microbiol. Immunol., 23: 361-367.

BENCKO, V., PEKAREK, J., SVEJCAR, J., BENES, B., HOLUSA, R., HURYCH, J., & SYMON, K. (1979b) [A contribution to the diagnosis of experimental berylliosis in the rat.] Cesk. Hyg., 24: 109-115 (in Czech, with English summary).

BENCKO, V., VASILYEVA, E.V., & SYMON, K. (1980) Immunological aspects of exposure to emissions from burning coal of high beryllium content. Environ. Res., 22: 439-449.

BERKOVITZ, M. & ISRAEL, B. (1940) [Changes in the lungs in fluorine beryllium poisoning.] Klin. Med. (Moscow), 18: 117-122 (in Russian).

BEUS, A.A. (1966) Geochemistry of beryllium and genetic types of beryllium deposits. San Francisco, California, W.H. Freeman and Co., 401 pp.

BIANVENU, P.K., NOFRE, C., & CIER, A. (1963) Toxicité générale comparée des ions métalliques. Relation avec la classification périodique. C. R. Séances Acad. Sci., 256: 1043.

BOBRISCHEV-PUSHKIN, D.M., NAUMOVA, L.A., & SURIKOVA, S.P. (1973) [Possible pollution of the industrial atmosphere with beryllium compounds in smelting BEB-2 bronze.] Gig. Tr. prof. Zabol., 6: 8-10 (in Russian, with English summary).

BOBRISCHEV-PUSHKIN, D.M., NAUMOVA, L.A., GRINBERG, A.A. & KHELKOVSKY-SERGEYEV, N.A. (1975) [Detection of different beryllium compounds at different types of welding.] Gig. Tr. prof. Zabol., 2: 41-43 (in Russian, with English summary).

BOBRISCHEV-PUSHKIN, D.M., NAUMOVA, L.A., KHELKOVSKY-SERGEYEV, N.A., & GRINBERG, A.A. (1976) [Health measures in improving conditions of work and environmental protection in production of metal beryllium articles.] Gig. Tr. prof. Zabol., 11: 6-9 (in Russian, with English summary).

BOIANO, J.M. (1980) Technical assistance report: Industrial hygiene report on a dental laboratory. Cincinnati, Ohio, National Institute for Occupational Safety and Health, Center for Disease Control (NIOSH Report No. 80-60-756).

BOKOWSKI, D.L. (1968) Rapid determination of beryllium by a direct-reading atomic absorption spectrophotometer. Am. Ind. Hyg. Assoc. J., 29: 474-481.

BOLAND, L.F. (1958) Beryllium. Present and potential uses. J. Met., June: 401-404.

BOWEN, H.J.M. (1966) Trace elements in biochemistry. London, New York, Academic Press, pp. 150-176.

BRESLIN, A.J. & HARRIS, W.B. (1959) Health protection in beryllium facilities. Arch. ind. Health, 19: 596-648.

BRINGMANN, V.G. & KUEHN, R. (1977) [Results of toxic action of water pollutants on *Daphnia magna*.] Z. Wasser Abwasser-Forsch., 10: 161-166 (in German).

BRINGMANN, V. G. & KUEHN, R. (1981) [Comparison of the effect of harmful substances on flagellates and ciliates as well as on bacteriovorous and saprozoic protozoans.] GWC-Wasser/Abwasser, 122: 308-310 (in German).

BRUMSACK, H., HEINRICHS, H., & LANGE, H. (1984) West German coal power plants as sources of potentially toxic emissions. Environ. Technol. Lett., 5: 7-22.

BUGRYSHEV, P.F., MOSKALEV, Y.I., KUZNETSOV, A.V., & NAZAROVA, A.A. (1976) [Beryllium distribution in rats.] Farmakol. Toksikol., 39: 615-618 (in Russian).

BURESCH, F.E. (1983) [Beryllia, thoria and zirconia ceramics.] Radex-Rundsch., 1/2: 133-145 (in German).

BURNAZIAN, A.I., ed. (1983) [Clinical picture of acute and chronic beryllium disorders.] Moscow, Meditsina, 176 pp. (in Russian).

BUSSY, M. (1828) Section de pharmacie: Glucinium. J. Chim. méd. Pharm. Toxicol., 4: 453-456.

BYRNE, C.J., BALASUBRAMANIAN, R., OVERTON, E.B., & ALBERT, T.F. (1985) Concentrations of trace metals in the bowhead whale. Mar. Pollut. Bull., 16: 497-498.

BYRNE, C.J. & DELEON, L.R. (1986) Trace metal residues in biota and sediments from Lake Pontchartrain, Louisiana. Bull. environ. Contam. Toxicol., 37(1): 151-158.

CAMNER, P., LUNDBORG, M., & HELLSTROM, P.A. (1974) Alveolar macrophages and 5 μm particles coated with different metals. Arch. environ. Health, 29(4): 211-213.

CAMNER, P., HELLSTROM, P.A., LUNDBORG, M., & PHILIPSON, K. (1977) Lung clearance of 4 μm particles coated with silver, carbon or beryllium. Arch. environ. Health, 32: 58-62.

CARDWELL, R.D., FOREMAN, D.G., PAYNE, T.R., & WILBUR, D.J. (1976) Acute toxicity of selected toxicants to six species of fish. Duluth Minnesota, US Environmental Protection Agency (Report No. 600/3-76-008).

CAROLI, S., CONI, E., ALIMONTI, A., BECCALONI, E., SABBIONI, E., & PIETRA, R. (1988) Determination of trace elements in human lungs by ICP-AES and NAA. Analusis., 16: 656-661.

CHAMBERS, L.S., FOTER, M.S., & CHOLAK, J. (1955) A comparison of particulate readings in the atmospheres of certain cities. In: Proceedings of the 3rd National Air Pollution Symposium, Pasadena, 18-20 April, 1955, pp. 24-32.

CHERNIACK, M.G. & KOMINSKY, J.R. (1984) Health hazard evaluation report. Chemetco, Incorporated, Alton, Illinois, Cincinnati, Ohio, National Institute for Occupational Safety and Health (NIOSH Report No. 82-024-1428).

CHEVREMONT, M. & FIRKET, H. (1951) Action of beryllium on cells cultivated *in vitro*; effect on mitosis. Nature (Lond.), 16: 772.

CHIAPPINO, G., CIVLA, A., & VIGLIANI, E.C. (1969) Delayed-type hypersensitivity reactions to beryllium compounds. Arch. Pathol., 87: 131-140.

CHOLAK, J., SCHAFER, L., & YEAGER, D. (1967) Exposures to beryllium in a beryllium alloying plant. Am. Ind. Hyg. Assoc. J., 28: 399-407.

CIKRT, M. & BENCKO, V. (1975) Biliary excretion of ^{7}Be and its distribution after intravenous administration of ^{7}BeCl$_2$ in rats. Arch. Toxicol., 34: 53-60.

CLARY, J.J., HOPPER, C.R., & STOKINGER, H.E. (1972) Altered adrenal function as an inducer of latent chronic beryllium disease. Toxicol. appl. Pharmacol., 23: 365-375.

CLOUDMAN, A.M., VINNING, D., BARKULIS, S., & NICKSON, J.J. (1949) Bone changes observed following intravenous injections of beryllium. Am. J. Pathol., 25: 810-811.

COCHRAN, K.W., ZERWIC, M.M., & DU BOIS, K.P. (1951) Studies on the mechanism of acute beryllium poisoning. J. Pharmacol. exp. Ther., 102: 165-178.

CONRADI, C., BURRI, P.H., KAPANCI, Y., ROBINSON, C.F.R., & WEIBEL, E.R. (1971) Lung changes after beryllium inhalation. Arch. environ. Health, 23: 348-358.

CONSTANTINIDIS, K. (1978) Acute and chronic beryllium disease. Br. J. clin. Pract., 32: 127-136, 153.

COTES, J.E., GILSON, J.C., MCKERROW, C.B., & OLDHAM, P.D. (1983) A long-term follow-up of workers exposed to beryllium. Br. J. ind. Med., 40(1): 13-21.

References

COVINGTON, J.S., MCBRIDE, M.A., SLAGLE, W.F., & DISNEY, A.L. (1985a) Beryllium localization in base metal dental casting alloys. J. biomed. Mater. Res., 19(7): 747-750.

COVINGTON, J.S., MCBRIDE, M.A., SLAGLE, W.F., & DISNEY, A.L. (1985b) Quantization of nickel and beryllium leakage from base metal casting alloys. J. Prostet. Dent., 5: 127-136.

CREMERS, D.A. & RADZIEMSKI, L.J. (1985) Direct detection of beryllium on filters using the laser spark. Appl. Spectrosc., 39(1): 57-63.

CROSSMON, G.C. & VANDEMARK, W.C. (1954) Microscopic observations correlating toxicity of beryllium oxide with crystal structure. Arch. ind. Health, 9: 481-487.

CROWLEY, J.F., HAMILTON, J.G., & SCOTT, K.G. (1949) The metabolism of carrier-free radioberyllium in the rat. J. biol. Chem., 177: 975-984.

CULLEN, M.R., CHERNIACK, M.G., & KOMINSKY, J.R. (1986) Chronic beryllium disease in the United States. Semin. respir. Med., 7: 203-209.

CULLEN, M.R., KOMINSKY, J.R., ROSSMAN, M.D., CHERNIACK, M.G., RANKIN, J.A., BALMES, J.R., KERN, J.A., DANIELE, R.P., PALMER, L., NAEGEL, G.P., McMAGNUS, K., & CRUZ, R. (1987) Chronic beryllium disease in a precious metal refinery. Clinical epidemiologic and immunologic evidence for continuing risk from exposure to low level beryllium fume. Am Rev. respir. Dis., 135: 201-208.

CUMMINGS, B., KASER, M.R., WIGGINS, G., ORD, M.G., & STOCKEN, L.A. (1982) Beryllium toxicity. The selective inhibition of casein kinase 1. Biochem. J., 208: 141-146.

CURTIN, G.C., KING, H.D., & MOSIER, E.L. (1974) Movement of elements into the atmosphere from coniferous trees in subalpine forests of Colorado and Idaho. J. geochem. Explor., 3: 245-263.

CURTIS, G.H. (1951) Cutaneous hypersensitivity due to beryllium. Arch. Dermatol. Syphilol., 64: 470-482.

CURTIS, G.H. (1959) The diagnosis of beryllium disease with special reference to the patch test. Arch. ind. Health, 19: 150-153.

DAVIS, R.D., BECKETT, P.H.T., & WOLLAN, E. (1978) Critical levels of twenty potentially toxic elements in young spring barley. Plant Soil, 49: 395-408.

DELVES, H.T. (1981) The analysis of biological and clinical materials. Prog. anal. at. Spectrosc., 4: 1-48.

DE NARDI, J.M., VAN ORDSTRAND, H.S., CURTIS, C.H., & ZIELINSKI, J. (1953) Berylliosis, summary and survey of all clinical types observed in a twelve-year period. Arch. ind. Hyg. occup. Med., 8(1): 10-24.

DENHAM, S. & HALL, J.G. (1988) Studies on the adjuvant action of beryllium. III. The activity in the plasma of lymph efferent from nodes stimulated with beryllium. Immunology, 64: 341-344.

DEODHAR, S.D., BARNA, B., & VAN ORDSTRAND, H.S. (1973) A study of the immunologic aspects of chronic berylliosis. Chest, 63: 309-313.

DFG (1988) [Maximum concentrations at the workplace and biological tolerance values for working materials 1988.] DFG, Deutsche Forschungsgemeinschaft, Weinheim, VCH Publishers, 99 pp. (Report No. 24 of the Commission for the investigation of health hazards of chemical compounds in the work area) (in German).

DILLING, W.J. & HEALEY, C.W. (1926) Influence of lead and the metallic ions of copper, zinc, thorium, beryllium, and thallium on the germination of frogs' spawn and on the growth of tadpoles. Ann. appl. Biol., 13: 177-188.

DITTMANN, J., HOFFEL, I., MULLER, P., & NEUHOEFFER, O. (1984) Use of poplar leaves for the monitoring of environmental beryllium. Naturwissenschaften, 71: 378.

DONALDSON, H.M. (1971) Results of air sampling of Kawechi Berlco plant. Cincinnati, Ohio, National Institute for Occupational Safety and Health, 30 pp (Unpublished report).

DONALDSON, H.M. & SHULER, P.J. (1972) Industrial hygiene survey of the Brush-Wellman plant, June 12-16, 1972, August 21-25, 1972. Cincinnati, Ohio, National Institute for Occupational Safety and Health, 65 pp (Unpublished report).

DREHER, G.B., MUCHMORE, C.B., & STOVER, D.W. (1977) Major, minor, and trace elements of bottom sediments in lake Du Quoin, Johnston City Lake, and Little Grassy Lake in Southern Illinois. Urbana, Illinois State Geological Survey, 38 pp (Environmental Geology Notes No. 82).

DRURY, J.S., SHRINER, C.R., LEWIS, E.G., TOWILL, L.E., & HAMMONS, A.S. (1978) Reviews of the environmental effects of pollutants: VI. Beryllium. Cincinatti, Ohio, US Environmental Protection Agency, 198 pp (Report No. EPA-600/1-78-028).

DUBOIS, K.P., COCHRAN, K.W., & MAZUR, M. (1949) Inhibition of phosphatases by beryllium and antagonism of the inhibition by manganese. Science, 110: 420-422.

DUDLEY, R.H. (1959) The pathologic changes of chronic beryllium disease. Arch. ind. Health, 19: 184-187.

DURUM, W.H. & HAFFTY, J. (1961) Occurrence of minor elements in water, Washington, DC, 11 pp (US Geological Survey Circular No.445).

DUTRA, F.R. (1948) Pneumonitis and granulomatosis peculiar to beryllium workers. Am. J. Pathol., 24: 1137-1165.

DUTRA, F.R. (1951) Experimental beryllium granulomas of the skin. Arch. ind. Hyg. occup. Med., 3: 81-89.

DUTRA, F.R. & LARGENT, E.J. (1950) Osteosarcoma induced by beryllium oxide. Am. J. Pathol., 26: 197-209.

DUTRA, F.R., LARGENT, E.J., & ROTH, J.L. (1951) Osteogenic sarcoma after inhalation of beryllium oxide. Arch. Pathol., 51: 473-479.

DVIVEDI, N. & SHEN, G. (1983) Beryllium toxicity in the laboratory processing of dental alloy. J. dent. Res., 62: 232.

EISENBUD, M. (1982) Origins of the standards for control of beryllium disease (1947-1949). Environ. Res., **27**: 79-88.

EISENBUD, M. & LISSON, J. (1983) Epidemiological aspects of beryllium-induced nonmalignant lung disease: a 30-year update. J. occup. Med., **25**(3): 196-202.

EISENBUD, M., BERGHOUT, C.F., & STEADMAN, L.T. (1948) Environmental studies in plants and laboratories using beryllium; the acute disease. J. ind. Hyg. Toxicol., **30**: 281-285.

EISENBUD, M., WANTA, R.C., DASTAN, C., STEADMAN, L.T., HARRIS, W.B., & WOLFE, B.S. (1949) Non-occupational berylliosis. J. ind. Hyg. Toxicol., **31**: 282-294.

ENCINA, C.L. & BECERRA, J. (1986) Inhibition of plant cytokinesis by beryllium and its revision by calcium. Environ. exp. Bot., **26**: 75-80.

EPSTEIN, M.S., REINS, T.C., BRADY, T.J., MOODY, J.R., & BARNES, I.L. (1978) Determination of several trace metals in simulated fresh water by graphite furnace atomic emission spectrometry. Anal. Chem., **50**: 874-880.

EPSTEIN, W.L. (1967) Granulomatous hypersensitivity. Prog. Allergy, **11**: 36-88.

ERBACH, G. (1984a) The incineration plant for special wastes in Biebesheim: Documentation. In: Proceedings of the 6th European Sewage and Refuse Symposium, Munich, 21-25 May, 1984. London, International Solid Wastes and Public Cleansing Association, pp. 629-642.

ERBACH, G. (1984b) [Industrial refuse disposal plant in Hesse (HIM) - experiences of the first year of operation.] VGB Kraftwerkstech., **64**(11): 1015-1019 (in German).

FARKAS, M.S. (1977) Beryllium - demand returns to normal levels in '76. Eng. Min. J., **178**: 169.

FINCH, G.L., MEWHINNEY, J.A., HOOVER, M.D., EIDSON, S.F., HALEY, P.J., & BICE, D.E. (1986) Toxicokinetics of beryllium following acute inhalation of BeO by beagle dogs. In: Muggenburg, B.A. & Sun, J.D., ed. Annual report. Inhalation Toxicology Research Institute, Albuquerque, Lovelace Biomedical and Environmental Research Institute, pp. 146-153.

FISHBEIN, L. (1981) Sources transport and alterations of metal compounds: an overview. 1. Arsenic, beryllium, cadmium, chromium, and nickel. Environ. Health Perspect., **40**: 43-64.

FISHBEIN, L. (1984) Overview of analysis of carcinogenic and/or mutagenic metals in biological and environmental samples. I. Arsenic, beryllium, cadmium, chromium and selenium. Int. J. environ. anal. Chem., **17**: 113-170.

FRANKE, W. (1985) [Useful plants.] Stuttgart, Thieme Verlag, 470 pp (in German).

FREIMAN, D.G. (1959) Pathologic changes of beryllium disease. Am. Med. Assoc. Arch. ind. Health, **19**: 188-189.

FREIMAN, D.G. & HARDY, H.L. (1970) Beryllium disease: The relation to pulmonary pathology to clinical course and prognosis based on a study of 130 cases from the U.S. Beryllium Case Registry. Hum. Pathol., 1(1): 25-44.

FREISE, R. & ISRAEL, G.W. (1987) [Investigations into the suspended dust load in Berlin (West)]. Berlin (West), Institut für technischen Umweltschutz, Technische Universität Berlin (in German).

FURCHNER, J.E., RICHMOND, C.R., & LONDON, J.E. (1973) Comparative metabolism of radionuclides in mammals. Part 8: Retention of beryllium in the mouse, rat, monkey and dog. Health Phys., 24(3): 293-300.

GARDNER, L.U. & HESLINGTON, H.F. (1946) Osteosarcoma from intravenous beryllium compounds in rabbits. Fed. Proc., 5: 221.

GELMAN, I. (1936) Poisoning by vapors of beryllium oxyfluoride. J. ind Hyg. Toxicol., 18: 371-379.

GELMAN, I. (1938) Beryllium (glucinium).In: ILOoccupation and health (Suppl.). Geneva, International Labour Office, pp.1-6.

GILLES, D. (1976) Health hazard evaluation determination. Hardric Laboratories, Waltham, Massachusetts, Cincinnati, Ohio, National Institute for Occupational Safety and Health (NIOSH Report No. 76-103-349).

GORMLEY, C.J. & LONDON, S.A. (1973) Effect of beryllium on soil microorganisms. In: Proceedings of the 4th Annual Conference on Environmental Toxicology, Fairborn, Ohio, 16-18 October 1973, Ohio, Wright-Patterson AFB, Aerospace Medical Research Laboratory, pp. 401-416 (AMRL-TR-73-125, Paper No. 29).

GREENFIELD, P. (1971) Engineering applications of beryllium. London, Mills & Boon Ltd, 52 pp (M. & B. Monograph ME/7).

GREWAL, D.S. & KEARNS, F.X. (1977) A simple and rapid determination of small amounts of beryllium in urine by flameless atomic absorption. At. Absorpt. Newsl., 16: 131-132.

GRIFFITTS, W.R., ALLAWAY, W.H., & GROTH, D.H. (1977) Beryllium. In: Geochemistry and the environment. Vol. II. The relation of other selected trace elements to health and disease. Washington, DC, National Academy of Sciences, US National Committee for Geochemistry, pp. 7-10.

GRIGGS, K. (1973) Toxic metal fumes from mantle-type camp lanterns. Science, 181(4102): 842-843.

GROTH, D.H. (1980) Carcinogenicity of beryllium: Review of the literature. Environ. Res., 21: 56-62.

GROTH, D.H., KOMMINENI, C., & MACKAY, G.R. (1980) Carcinogenicity of beryllium hydroxide and alloys. Environ. Res., 21: 63-84.

GUNTER, B.J. & THOBURN, T.W. (1986) Health hazard evaluation report. Rockwell International, Rocky Flats Plant, Golden, Colorado, Cincinnati, Ohio, National Institute for Occupational Safety and Health, Center for Disease Control (NIOSH Report No. 84-510-1691).

GUYATT, B.L., KAY, H.D., & BRANION, H.D. (1933) Beryllium "rickets". J. Nutr., 6: 313-324.

HALL, J.G. (1988) Studies on the adjuvant action of beryllium. IV. The preparation of beryllium containing macromolecules that induce immunoblast responses in vivo. Immunology, 64: 345-351.

HALL, R.H., SCOTT, J.K., LASKIN, S., STROUD, C.A., & STOKINGER, H.E. (1950) Acute toxicity of inhaled beryllium. III. Observations correlating toxicity with the physicochemical properties of beryllium oxide dust. Arch. ind. Hyg. occup. Med., 2: 25-33.

HALL, T.C., WOOD, C.H., STOECKLE, J.D., & TEPPER, L.B. (1959) Case data from the beryllium registry. Am. Med. Assoc. Arch. ind. Health, 19: 100-103.

HAMILTON, E.I. & MINSKY, M.J. (1973) Abundance of the chemical elements in man's diet and possible relations with environmental factors. Sci. total Environ., 1: 375-394.

HARA, T., SONODA, Y., & IWAI, I. (1977) Growth response of cabbage plants to beryllium and strontium under water culture conditions. Soil Sci. plant Nutr., 23: 373-380.

HARDY, H.L. (1948) Delayed chemical pneumonitis in workers exposed to beryllium compounds. Am. Rev. Tuberc., 57: 547-556.

HARDY, H.L. (1965) Beryllium poisoning - lessons in control of man-made disease. New Engl. J. Med., 273: 1188-1199.

HARDY, H.L. (1980) Beryllium disease: A clinical perspective. Environ. Res., 21: 1-9.

HARDY, H.L. & CHAMBERLIN, R.I. (1972) Beryllium disease. In: Tabershaw, I.R., ed. The toxicology of beryllium. Washington, DC, US Department of Health Education and Welfare, Public Health Service, pp. 5-16 (Publication No. 82-214230).

HARDY, H.L. & STOECKLE, J.D. (1959) Beryllium disease. J. chron. Dis., 9: 152-160.

HARDY, H.L. & TABERSHAW, I.R. (1946) Delayed chemical pneumonitis occurring in workers exposed to beryllium compounds. J. ind. Hyg., 28: 197-211.

HARDY, H.L., RABE, E.W., & LURCH, S. (1967) United States beryllium case registry (1952-1966): Review of its methods and utility. J. occup. Med., 9: 271-276.

HART, B.A. & PITTMAN, D.G. (1980) The uptake of beryllium by the alveolar macrophage. J. Reticuloendothel. Soc., 27: 49-58.

HART, B.A., HARMSEN, A.G., LOW, R.B., & EMERSON, R. (1984) Biochemical, cytological, and histological alterations in rat lung following acute beryllium aerosol exposure. Toxicol. appl. Pharmacol., 75: 454-465.

HILDEBRAND, S.G. & CUSHMAN, R.M. (1978) Toxicity of gallium and beryllium to developing carp eggs (*Cyprinus carpio*) utilizing copper as a reference. Toxicol. Lett., 2(2): 91-95.

HOAGLAND, M.B. (1952a) Beryllium and growth. II. The effect of beryllium on plant growth. Arch. Biochem. Biophys., 35: 249-258.

HOAGLAND, M.B. (1952b) Beryllium and growth. III. The effect of beryllium on plant phosphatase. Arch. Biochem. Biophys., 35: 259-267.

HOAGLAND, M.B., GRIER, R.S., & HOOD, M.B. (1950) Beryllium-induced osteogenic sarcomata. Cancer Res., 10: 629-635.

HOGBERG, T. & RAJS, J. (1980) Two cases of sudden death due to granulomatous myocarditis among beryllium exposed workers. Toxicol. Lett., 5: 203.

HOKIN, L.E., HOKIN, M.R., & MATHISON, D. (1963) Phosphatidic acid phosphatase in the erythrocyte membrane. Biochim. biophys. Acta, 67: 485-497.

HOOPER, W.F. (1981) Acute beryllium lung disease. North Carolina med. J., 42: 551-553.

HSIE, A.W., O'NEILL, J.P., SAN SABASTIAN, J.R., COUCH, D.B., BRIMER, P.A., SUN, W.N.C., FUSCOE, J.C., FORBES, N.L., MACHANOFF, R., RIDDLE, J.C., & HSIE, M.H. (1979a) Quantitative mammalian cell genetic mutagenicity of seventy individual environmental agents related to energy technology and three subfractions of crude synthetic oil in the CHO/HGPRT system. Environ. Sci. Res., 15: 291-315.

HSIE, A.W., JOHNSON, M.P., COUCH, D.B., SAN SABASTIAN, J.R., O'NEILL, J.P., HOESCHELE, J.D., RAHN, R.O., & FORBES, N.L. (1979b) Quantitative mammalian cell mutagenesis and a preliminary study of the mutagenic potential of metallic compounds. In: Kharsch, N., ed. Trace metals in health and disease. New York, Raven Press, pp. 55-69.

HURLBUT, J.A. (1978) Determination of beryllium in biological tissues and fluids by flameless atomic absorption spectroscopy. At. Absorpt. Newsl., 17: 121-124.

HYSLOP, F., PALMES, E.D., ALFORD, W.C., MONACO, A.R., & FAIRHALL, L.T. (1943) The toxicology of beryllium. Natl Inst. Health Bull., 181: 1-56.

IARC (1972) Beryllium and beryllium compounds. In: Inorganic substances. Lyon, International Agency for Research on Cancer, pp. 17-28 (Monographs on the Evaluation of the Carcinogenic Risk of Chemicals to Humans, Vol. 1).

IARC (1980) Beryllium and beryllium compounds. In: Some metals and metallic compounds. Lyon, International Agency for Research on Cancer, pp. 143-204 (Monographs on the Evaluation of the Carcinogenic Risk of Chemicals to Humans, Vol. 23).

IARC (1986) Determination of beryllium in air by graphite-furnace atomic absorption spectrophotometry. In: O'Neill, I.K., Schuller, P., & Fishbein, L., ed. Environmental carcinogens: Selected methods of analysis. Lyon, International Agency for Research on Cancer, pp. 283-288 (IARC Scientific Publications No. 71).

IARC (1987) Overall evaluations of carcinogenicity: An updating of IARC Monographs, Volumes 1 to 42. Lyon, International Agency for Research on Cancer, pp. 127-128 (Supplement 7 to IARC Monographs on the Evaluation of Carcinogenic Risks to Humans).

IAWR (1986) [Report of the international working group of the water works in the Rhine River area 1983-1985.] Amsterdam, International Arbeitsgemeinschaft der Wasserwerke im Rheineinzubsgebiet, 84 pp (in German).

ICRP (1960) Report of ICRP Committee II on Permissible Dose for Internal Radiation. Health Phys., 3: 154-155.

IKEBE, K., TANAKA, R., KUZUHARA, Y., SUENAGA, S., & TAKABATAKE, E. (1986) Studies on the behavior of beryllium in environment. Behavior of beryllium and strontium in atmospheric air. Eisei Kagaku, 32: 159-166.

INFANTE, P.F., WAGONER, J.K., & SPRINCE, N.L. (1980) Mortality patterns from lung cancer and nonneoplastic respiratory disease among white males in the beryllium case registry. Environ. Res., 21: 35-43.

ISHINISHI, N., MIZUNOE, M., INAMASU, T., & HISANAGA, A. (1980) [Experimental study of the carcinogenicity of beryllium oxide and arsenic trioxide to the lung of the rat by intratracheal instillation.] Fukuoka Igaku Zasshi, 71(1): 19-26 (in Japanese).

IZMEROV, N.F. (ed.) (1985) Beryllium. Series "Scientific Reviews of Soviet Literature on Toxicity and Hazards of Chemicals", International Register of Potentially Toxic Chemicals and Centre of International Projects, Moscow, GKNT, 49 pp.

IZUMI, T., KOBARA, Y., INUI, S., TOKUNAGA, R., ORITA, Y., KITANO, M., & JONES WILLIAMS, W. (1976) The first seven cases of chronic beryllium disease in ceramic factory workers in Japan. Ann. NY Acad. Sci., 278: 636-653.

JANES, J.M., HIGGINS, G.M., & HERRICK, G.F. (1954) Beryllium-induced osteogenic sarcoma in rabbits. J. Bone Joint Surg. Br., 36B: 543-552.

JANES, J.M., HIGGINS, G.M., & HERRICK, G.F. (1956) The influence of splenectomy on the induction of osteogenic sarcoma in rabbits. J. Bone Joint Surg. Am., 38A: 809-816.

JIRELE, V., NECHYBA, L. & PACHNER, P. (1966) [A contribution on the elucidation of beryllium hygienic importance in some sorts of Sokolov Basin coal.] Cesk. Hyg., 11: 329-339 (in Czech, with English summary).

JONES WILLIAMS, W. (1958) A histological study of the lungs in 52 cases of chronic beryllium disease. Br. J. ind. Med., 15: 84-91.

JONES WILLIAMS, W. (1977) Beryllium disease, pathology and diagnosis. J. occup. Med., 27: 93-96.

JONES WILLIAMS, W. (1985) UK beryllium registry 1945-1985. J. Pathol., 146: 284.

JONES WILLIAMS, W. (1988) Beryllium disease. Postgrad. med. J., 64: 511-516.

JONES WILLIAMS, W. & KELLAND, D. (1986) New aid for diagnosing chronic beryllium disease: Laser ion mass analysis. J. clin. Pathol., 39: 900-901.

JONES WILLIAMS, W. & WILLIAMS, W.R. (1974) Cutaneous and pulmonary manifestations of chronic beryllium disease. In: Iwai, K. & Hosoda, Y., ed. Proceedings of the VI International Conference on Sarcoidosis, Tokyo. Tokyo, University of Tokyo Press, pp. 141-145.

JONES WILLIAMS, W. & WILLIAMS, W.R. (1983) Value of beryllium lymphocyte transformation tests in chronic beryllium disease and in potentially exposed workers. Thorax, 38(1): 41-44.

JONES WILLIAMS, W., FRY, E., & JAMES, E.M.V. (1972) The fine structure of beryllium granulomas. Acta Pathol. Microbiol. Scand. (Suppl.), 233: 195-202.

JONES WILLIAMS, W., WILLIAMS, W.R., KELLAND, D., & HOLT, P.J.A. (1988) Beryllium skin disease. In: Grassi, C., Rizzato, G., & Pozzi, E., ed., Sarcoidosis and other granulomatous disorders. Amsterdam, Oxford, New york, Elsevier Science Publishers, pp. 689-690.

KAISER, G., GRALLATH, E., TSCHOPEL, P., & TOLG, G. (1972) [Contribution to the optimization of the chelate gas-chromatographic determination of beryllium in limited amounts of organic materials.] Z. anal. Chem., 259: 257-264 (in German).

KANAREK D.J., WAINER R.A., CHAMBERLIN, R.I., WEBER, A.L., & KAZEMI, H. (1973) Respiratory illness in a population exposed to beryllium. Am. Rev. respir. Dis., 108(6): 1295-1302.

KANEMATSU, N., HARA, M., & KADA, T. (1980) Rec assay and mutagenicity studies on metal compounds. Mutat. Res., 77: 109-116.

KARLANDER, E.P. & KRAUSS, R.W. (1972) Absorption and toxicity of beryllium and lithium in *Chlorella vannielii* Shihira and Krauss. Chesapeake Sci., 13: 245-253.

KATZ, S.A. (1985) Collection and preparation of biological tissues and fluids for trace element analysis. Int. Biotechnol. Lab., 3: 10-16.

KEENAN, R.C. & HOLTZ, J.L. (1964) Spectrographic determination of beryllium in air, biological materials, and ores using the sustaining A.C. arc. Am. Ind. Hyg. Assoc. J., 25: 254-263.

KELLEY, W.N., GOLDFINGER, S.E., & HARDY, H.L. (1969) Hyperuricemia in chronic beryllium disease. Ann. int. Med., 70: 977-983.

KELLY, P.J., JANES, J.A., & PETERSON, L.F.A. (1961) The effect of beryllium on bone. J. Bone Joint Surg. Am., 43A: 829-844.

KICK, H., BURGER, H., & SOMMER, K. (1980) [Plant experiments on the uptake of beryllium and thallium by barley and rape.] Landwirtsch. Forsch., 37: 186-190 (in German).

KIMMERLE, G. (1966) [Beryllium.] In: [Handbook of experimental pharmacology.] Berlin, New york, Springer-Verlag, Vol. XXI, 80 pp. (in German).

KINGSTON, H.M. & JASSIE, L.B. (1986) Microwave energy for acid decomposition at elevated temperatures and pressures using biological and botanical samples. Anal. Chem., 58: 2534-2541.

KJELLSTROM, T. & KENNEDY, P. (1984) Criteria document for Swedish occupational standards: Beryllium. Solna, Sweden, Arbetarskyddsstyrelsen, Publikationsservice, 60 pp.

KLOKE, A., SAUERBECK, D.R., & VETTER, H. (1984) The contamination of plants and soils with heavy metals and the transport of metals in terrestrial food chains. In. Nriagu, J.O., ed. Changing metal cycles and human health. Report of the Dahlem Workshop, Berlin (FRG), 20-25 March, 1983. Berlin, New York, Springer-Verlag, pp. 113-141.

KOMITOWSKI, D. (1967) [Experimental beryllium conducted bone tumours as a model of osteogenic sarcoma.] Chir. Narzadow Ruchu Ortop. Pol., **33**: 237-242 (in Polish).

KRAMPITZ, G. (1980) [Beryllium.] Bild Wiss., **11**: 154-161 (in German).

KRAMPITZ, G., HATJIPANAGIOTOU, A., LESUR, E., HARDEBECK, H., & GREUEL, E. (1978) [Effects of beryllium compounds on the hen. Part. 1. Toxicity of beryllium sulfate in the hen.] Arch. Geflügelkd., **43**: 225-228 (in German).

KREISS, K., NEWMAN, L.S., MROZ, M.M. & CAMPELL, P.A. (1989) Screening blood test identifies subclinical beryllium disease. J. occup. Med., **31**: 603-608.

KREJCI, L.E. & SCHEEL, L.D. (1966) The chemistry of beryllium. In: Stokinger, H.E., ed. Beryllium - its industrial hygiene aspects. New York, London, Academic Press, pp. 45-131.

KRESS, J.E. & CRISPELL, K.R. (1944) Chemical pneumonitis in men working with fluorescent powders containing beryllium. Guthrie clin. Bull., **13**: 91-95.

KRIEBEL, D., SPRINCE, N.L., EISEN, E.A., GREAVES, I.A., FELDMAN, H.A., & GREENE, R.E. (1988) Beryllium exposure and pulmonary function: a cross sectional study of beryllium workers. Br. J. ind. Med., **45**: 167-173

KRIVANEK, N.D. & REEVES, A.L. (1972) The effect of chemical forms of beryllium on the production of the immunologic response. Am Ind. Hyg. Assoc. J., **33**: 45-52.

KRIVORUTCHKO, F.D. (1966) [Photometric method of beryllium detecting in the air with phosphon-azo R.] Gig. i Sanit., **4**: 57-60 (in Russian).

KUBINSKI, H., GUTZKE, G.E., & KUBINSKIE, Z.O. (1981) DNA-cell-binding (DCS) assay for suspected carcinogens and mutagens. Mutat. Res., **89**: 95-136.

KUBINSKI, P.E., ZELDIN, P.E., & MORIN, N.R. (1977) Survey of tumor-producing agents for their ability to induce macromolecular complexes. Proc. Am. Assoc. Cancer Res., **18**: 16.

KUSCHNER, M. (1981) The carcinogenicity of beryllium. Environ. Health Perspect., **40**: 101-106.

KUZNETSOV, A.V., MATVEER, O.G., & SUNTSOC, G.D. (1974) [Differences in the distribution of labelled beryllium chloride with or without carrier in rats following intratracheal administration.] Gig. i Sanit., **10**: 113-114 (in Russian).

KWAPULINSKI, J. & PASTUSZKA, J. (1983) Application of the mass balance equation in the estimation of beryllium and radium concentrations in the lower atmosphere. Sci. total Environ., **26**: 203-207.

LANGHANS, D. (1984) [The influence of beryllium on the germination of garden cress (*Lepidium sativum* L.).] Angew. Bot., **58**: 295-299 (in German).

LARRAMENDY, M.L., POPESCU, N.C., & DIPAOLO, J.A. (1981) Induction by inorganic metal salts of sister chromatid exchanges and chromosome aberrations in human and Syrian hamster cell strains. Environ. Mutagen., 3: 597-606.

LASKIN, S.R., TURNER, A.N., & STOKINGER, H.E. (1950) An analysis of dust and fume hazards in a beryllium plant. In: Vorwald, A.J., ed. Pneumoconiosis: Beryllium and bauxite fumes. New York, Hoeber Inc., pp. 360-386.

LEBEDENA, G.D. (1960) [Effect of beryllium chloride on aquatic organisms.] Zool. a. Zh., 39: 1779-1782 (in Russian).

LEDERER, H. & SAVAGE, J. (1954) Beryllium granuloma of the skin. Br. J. ind. Med., 11: 45-51.

LEWIS, F.A. (1980) Health hazard evaluation determination report. Bertoia Studio, Bally, Pennsylvania, Cincinnati, Ohio, National Institute for Occupational Safety and Health (NIOSH Report No. 79-78-655).

LIEBEN, J. & WILLIAMS, R.R. (1969) Respiratory disease associated with beryllium refining and alloy fabrication: 1968 follow-up. J. occup. Med., 11: 480-485.

LIEBEN, J., DATTOLI, J.A., & VOUGHT, V.M. (1966) The significance of beryllium concentrations in urine. Arch. environ. Health, 12: 331-334.

LITVINOV, N.N., POPOV, V.A., VOROZHEIKINA, T.V., KAZENASHEV, V.F., & BUGRYSHEV, P.F. (1984) [Data for more precise determination for the MAC for beryllium in the air of the workplace.] Gig. Tr. prof. Sanit., 1: 34-37 (in Russian, with English summary).

LOVBLAD, G. (1977) Trace elements concentrations in some coal samples and possible emissions from coal combustion in Sweden. Gothenburg, Swedish Water and Air Pollution Research Laboratory, 20 pp.

LUKE, M.Z., HAMILTON, L., & HOLLOCHER, T.C. (1975) Beryllium-induced misincorporation by a DNA polymerase. Possible factor in beryllium toxicity. Biochem. biophys. Res. Commun., 62(2): 497-501.

LUNDBORG, M., LIND, B., & CAMNER, P. (1984) Ability of rabbit alveolar macrophages to dissolve metals. Exp. lung Res., 7: 11-22.

LUPTON, D. & ALDINGER, F. (1983) [Beryllium - a light metal with special properties - and its handling.] Radex-Rundsch., 1/2: 43-51 (in German).

McCANN, J.E., CHOI, E., YAMASAKI, E., & AMES, B.N. (1975) Detection of carcinogens as mutagens in the $Salmonella$/microsome test: Assay of 300 chemicals. Proc. Natl Acad. Sci. (USA), 72: 5135-5139.

McCORD, C.P. (1951) Beryllium as a sensitizing agent. Ind. Med. Surg., 20: 336.

MACCORDICK, J., YOUNINOU, M.-TH., & WURTZ, B. (1976) Effect of potassium dioxalatoberyllate, $K_2[Be(C_2O_4)_2]$, on the growth of $Pseudomonas$ $aeruginosa$. Naturwissenschaften, 63: 90-91.

MACMAHON, H.E. & OLKEN, H.G. (1950) Chronic pulmonary berylliosis in workers using fluorescent powders containing beryllium. Arch. ind. Hyg. occup. Med., 1: 195-214.

McMANUS, K.P., KOMINSKY, J.R., CHERNIACK, M.G., MCCONNELL, R., & HANDKE, J. (1986) Health hazard evaluation report. Handy and Harman, Incorporated, Fairfield, Connecticut, Cincinnati, Ohio, National Institute for Occupational Safety and Health, Center for Disease Control (NIOSH Report No. 83-162-1746).

MAINIGI, K.D. & BRESNICK, E. (1969) Inhibition of deoxythymidine kinase by beryllium. Biochem. Pharmacol., **18**: 2003-2007.

MANCUSO, T.F. (1970) Relation of duration of employment and prior repiratory illness to respiratory cancer among beryllium workers. Environ. Res., **3**: 251-275.

MANCUSO, T.F. (1979) Occupational lung cancer among beryllium workers: Dusts and disease. In: Lemen, R. & Dement, J., ed. Proceedings of the Conference on Occupational Exposure to Fibrous and Particulate Dust and their Extension into the Environment. Park Forest South, Illinois, Pathotox Publishers, pp. 463-471.

MANCUSO, T.F. (1980) Mortality study of beryllium industry workers occupational lung cancer. Environ. Res., **21**: 48-55.

MANCUSO, T.F. & EL-ATTAR, A.A. (1969) Epidemiologic study of the beryllium industry: Cohort methodology and mortality studies. J. occup. Med., **11**: 422-434.

MARRADI-FABRONI, S.M. (1935) Pulmonary pathology of beryllium powders. Med. Lav., **26**: 277-303.

MARSHALL, A.T. & CARDE, D. (1984) Beryllium coating for biological X-ray microanalysis. J. Microsc., **134**: 113-116.

MASON, B. (1952) Principles of geochemistry. Chichester, New York, John Wiley and Sons, 276 pp.

MATHUR, R., SHARMA, S., MATHUR, S., & PRAKASH, A.O. (1987) Effect of beryllium nitrate on early and late pregnancy in rats. Bull. environ. Contam. Toxicol., **38**: 73-77.

MEASURES, C.I. & EDMOND, J.M. (1982) Beryllium in the water column of the central North Pacific. Nature (Lond.), **297**: 51-53.

MEEHAN, W.R. & SMYTHE, L.E. (1967) Occurrence of beryllium as a trace element in environmental materials. Environ. Sci. Technol., **1**: 839-844.

MELNIKOV, V.V. (1959) [Materials on the toxicology of beryllium acetate.] Farmakol. Toksikol., **22**: 261-269 (in Russian).

MENESINI, G. (1938) [Biological action of beryllium carbonate.] Rass. Med. Lav. ind., **8**: 317-340 (in Italian).

MERRIL, J.R., LYDEN, E.F.X., HONDA, M., & ARNOLD, J.R. (1960) Sedimentary geochemistry of the beryllium isotopes. Geochim. cosmochim. Acta, **18**: 108-129.

MEYER, H.E. (1942) [On beryllium-caused lung diseases.] Beitr. klin. Tuberk., **98**: 388-395 (in German).

MINKWITZ, R., FROHLICH, N., & LEHMANN, E., (1983) [Examination of charges of harmful substances at work places during the production and processing of metals: beryllium, cobalt and their alloys.] Dortmund, Bundesanstalt für Arbeitsschutz, 107 pp (Forschungsbericht No. 367) (in German).

MIYAKI, M., AKAMATSU, N., ONO, T., & KOYAMA, H. (1979) Mutagenicity of metal cations in cultured cells from Chinese hamsters. Mutat. Res., 68: 259-263.

MORGAREIDGE, K., COX, G.E., BAYLEY, D.E., & GALLO, M. (1977) Chronic oral toxicity of beryllium in the rat. Toxicol. appl. Pharmacol., 41: 204-205.

MORIMOTO, F. (1959) [An experimental study of toxicosis by beryllium.] Fukuoka Acta med., 50: 4398-4430 (in Japanese).

MOSELEY, C.L. & DONOHUE, M.T. (1983) Health hazard evaluation report: International Brotherhood of Painters and Allied Trades, Electric Boat Division of General Dynamic Corporation, Cincinnati, Ohio, National Institute for Occupational Safety and Health (NIOSH Report No. 78-135-1333).

MUELLER, J. (1979) Beryllium, cobalt, chromium and nickel in particulate matter of ambient air. In: Proceedings of the International Conference of Heavy Metals in the Environment, London, September 1979. Edinburgh, CEP Consultants Ltd., pp. 300-303.

MUELLER, P., DMOWSKI, K., GAST, F., HAHN, E., & WAGNER, G. (1984) [Problem of bird feathers as location-specific bioindicators for heavy-metal loadings.] Wiss. Umwelt, 3: 139-144 (in German).

MUELLER, P., FLACKE, W., & HOEFFEL, I. (1986) [Development of a biomonitoring system for the Hoecherberg area.] Saarbrücken, Federal Republic of Germany, Institute for Biogeography. Umweltforschungs plan des Bundesministers des Innern (Research Report No. 109.02.001) (in German).

MUKHINA, S.T. (1967) [Effect of magnesium on oxidative processes in rat liver and lung homogenates as a result of experimental beryllium intoxication.] Gig. Tr. prof. Zabol., 11: 43-46 (in Russian).

MULLEN, A.L., STANLEY, R.E., LLOYD, S.R., & MOGHISSI, A.A. (1972) Radioberyllium metabolism by the dairy cow. Health Phys., 22(1): 17-22.

MULWANI, H.R. & SATHE, R.M. (1977) Spectrophotometric determination of beryllium in air by a sensitised chrome Azurol S reaction. Analyst, 102: 137-139.

NATIONAL ACADEMY OF SCIENCES (1977) Drinking water and health. Washington, DC, National Academy of Sciences, Safe Drinking Water Committee.

NAUMOVA, L.A. (1967) [Hygienic evaluation of the process producing aluminium-beryllium alloys,] Moscow, Institute of Labour, Hygiene and Professional Diseases, pp. 36-38 (in Russian).

NAUMOVA, L.A. & GRINBERG, A.A. (1974) [Detecting beryllium in the air.] Saf. Meas. Purif. sewage Syst. chem. Ind. (Moscow), 10: 15-19 (in Russian).

NEWMAN, L.S., KREISS, K., KING, T.E., SEAY, S., & CAMPBELL, P.A. (1988) Pathologic and immunologic alterations in early stages of beryllium disease. Am. Rev. respir. Dis., 139: 1479-1486

NEWLAND, L.W. (1982) Arsenic, beryllium, selenium and vanadium. In: Hutzinger, O., ed. The handbook of environmental chemistry. Berlin, New York, Springer-Verlag, Vol. 3, Part B, pp. 27-67.

NIKONOVA, N.N. (1967) [On the accumulation of beryllium, molybdenum, zirconium, yttrium, and other rare elements in plants of South Ural.] Izv. Sib. Akad. Nauk. SSSR Ser. Biol. Med., 3: 25-29 (in Russian).

NILSEN, A., NYBERG, M, K., & CAMNER, P. (1988) Intraphagosomal pH in alveolar macrophages after phagocytosis *in vivo* and *in vitro* of fluorescein-labelled yeast particles. Exp. lung Res., 14: 197-207.

NIOSH (1972) Criteria document: Recommendations for an occupational exposure standard for beryllium. Rockville, Maryland, National Institute for Occupational Safety and Health, 128 pp (NIOSH Report No. TR-003-72: HSM 72-10268).

NIOSH (1977) Manual of analytical methods. 2nd ed. Cincinnati, Ohio, National Institute for Occupational Safety and Health, Vol. 3, p. 339.

NIOSH (1984) Manual of analytical methods. 3rd ed. Cincinnati, Ohio, National Institute for Occupational Safety and Health, Vol. 1, pp. 71021-71023.

NISHIMURA, A. (1966) Clinical and experimental studies on acute beryllium disease. Nagoya J. med. Sci., 28: 17-44.

NISHIOKA, H. (1975) Mutagenic activities of metal compounds in bacteria. Mutat. Res., 31: 185-189.

OSHA (1975) Occupational exposure to beryllium. Notice of proposed rulemaking. Fed. Reg., 40(202): 48814-48827.

PARFENOV, Y.D. (1988) [Calculation of the maximal allowable concentration of beryllium in the air on the basis of carcinogenic effects.] Gig. i Sanit., 6: 59-62 (in Russian).

PARKER, V.H. & STEVENS, C. (1979) Binding of beryllium to nuclear acidic proteins. Chem. biol. Interact., 26: 167-177.

PATON, G.R. & ALLISON, A.C. (1972) Chromosome damage in human cell cultures induced by metal salts. Mutat. Res., 16: 332-336.

PETZOW, G. & ALDINGER, F. (1974) [Beryllium and beryllium compounds.] In: Bartholomé, E., Bickert, E., Hellmann, H., & Ley, H. ed. [Ullmanns encyclopedia of technical chemistry.] Weinheim, Verlag Chemie, pp. 442-458 (in German).

PETZOW, G. & ZORN, H. (1974) [Toxicology of beryllium containing materials.] Chem. Ztg, 98: 236-241 (in German).

PIRSCHLE, K. (1935) [Comparative experiments on physiological effects of the elements on *Aspergillus niger* (stimulation and toxicity).] Planta med., 23: 177-224 (in German).

POLICARD, A. (1948) Etude expérimentale sur l'action pulmonaire de poussières renfermant des composés de glucinium (beryllium). Bull. Acad. nat. Méd., 132: 449-451.

POLICARD, A. (1949a) Experimental research on the action of dust of beryllium compounds on the lung. In: Proceedings of the Ninth International Congress on Industrial Medicine, London, England, 13-17 September, 1948. Bristol, John Wright & Sons, pp.796-797.

POLICARD, A. (1949b) La bérylliose pulmonaire expérimentale chez la rat et ses mécanismes pathogéniques. J. fr. Méd. Chir. thorax, **6**: 501-517.

POLICARD, A. (1950a) Importance des éléments fluorée dans la pathogénie des pneumopathies par poussières des composés bérylliques. Bull. Acad. nat. Méd., **134**: 289-290.

POLICARD, A. (1950b) Histological studies of the effects of beryllium oxide (glucine) on animal tissues. Br. J. ind. Med., **7**: 117-121.

POWERS, P.W. (1976) How to dispose of toxic substances and industrial wastes. Park Ridge, New Jersey, Noyes Data Corporation.

PREUSS, O.P. (1975) Toxicity of beryllium and its compounds. Proceedings of the Symposium on the Toxicicity of Metals, Pittsburgh, 5-6 June, 1974. Pittsburgh, Pennsylvania Industrial Health Foundation, pp. 31-43.

PREUSS, O.P. (1985) Long term follow up of workers exposed to beryllium. Br. J. ind. Med., **42**: 69-72.

PREUSS, O.P. & OSTER, H. (1980) [Health risk from beryllium.] Arbeitsmed., Sozialmed., Präventivmed., **15**: 270-275 (in German).

PREUSS, O.P., DEODHAR, S.D., & VAN ORDSTRAND, H.S. (1980) Lymphoblast transformation in beryllium workers. In: Williams, W.J. & Davies, B.H., ed. Sarcoidosis and other granulomatous diseases. Cardiff, Alpha-Omega Publications, pp. 711-714.

PUGLIESE, P.T., WHITLOCK, L.M., ERNEST, B., & WILLIAMS, R.R. (1968) Serological reactions of berylliosis patients to extracts of lung tissue. Arch. environ. Health, **16**: 374-379.

PUZANOVA, L., DOSKOCIL, M., & DOUBKOVA, A. (1978) Distribution of the development of chick embryos after the administration of beryllium chloride at early stages of embryogenesis. Folia morphol. (Prague), **26**: 228-231.

REEVES, A.L. (1965) The absorption of beryllium from the gastrointestinal tract. Arch. environ Health, **11**: 209-214.

REEVES, A.L. (1968) [Retention of inhaled beryllium sulfate aerosol in rat lungs.] Int. Arch. Gewerbepath. Gewerbehyg., **24**: 226-237 (in German).

REEVES, A.L. (1977) Beryllium in the environment. Clin. Toxicol., **10**(1): 37-48.

REEVES, A.L. (1978) Beryllium carcinogenesis. In: Schranzer, G.N., ed. Inorganic and nutritional aspect of cancer. New York, London, Plenum Press, pp. 13-27.

REEVES, A.L. (1986) Beryllium. In: Friberg. L., Nordberg. G.F., & Vouk. V., ed. Handbook on the toxicology of metals. 2nd ed. Amsterdam, New York, Oxford, Elsevier, Science Publishers, Vol. II, pp. 95-116.

References

REEVES, A.L. & DEITCH, D. (1969) Influence of age on the carcinogenic response to beryllium inhalation. In: Harishima, S., ed. Proceedings of the 16th International Congress on Occupational Health, Tokyo. Tokyo, Japan Industrial Safety Association, pp. 651-652.

REEVES, A.L. & PREUSS, O.P. (1985) The immunotoxicity of beryllium. In: Dean, J. Luster, M.I., Munson, A.E., & Amos, H., ed. Immunotoxicity and immunopharmacology. New York, Raven Press, pp. 441-455.

REEVES, A.L. & PREUSS, O.P. (in press) The immunological diagnosis of chronic beryllium disease in humans and animals. In: Proceedings of the International Workshop on Immunotoxicity and Immunotoxicology of Metals, Hanover, Federal Republic of Germany, 6-10 November 1989. New York, London, Plenum Press.

REEVES, A.L. & VORWALD, A.J. (1967) Beryllium carcinogenesis. II. Pulmonary deposition and clearance of inhaled beryllium sulfate in the rat. Cancer Res., 27: 446-451.

REEVES, A.L., DEITCH, D., & VORWALD, A.J. (1967) Beryllium carcinogenesis. I. Inhalation exposure of rats to beryllium sulfate aerosol. Cancer Res., 27: 439-445.

REEVES, A.L., KRIVANEK, N.D., & BUSBY, E.K. (1972) Immunity to pulmonary berylliosis in guinea pigs. Int. Arch. Arbeitsmed., 29(3): 209-220.

REICHERT, J.K. (1974) [Beryllium: A toxic element in the human environment with special regard to its occurrence in water.] Vom Wasser, 41: 209-216 (in German).

REINER, E. (1971) Binding of beryllium to proteins. In: Aldridge, W.N., ed. Symposium on Mechanisms of Toxicity, London, 13-14 April, 1970. London, Macmillan Press Ltd, pp. 111-125.

RESNICK, H., ROCHE, M., & MORGAN, W.K.C. (1970) Immunoglobulin concentration in berylliosis. Am. Rev. resp. Dis., 101: 504-510.

RHOADS, K. & SANDERS, C.L. (1985) Lung clearance, translocation, and acute toxicity of arsenic, beryllium, cadmium, cobalt, lead, selenium, vanadium, and ytterbium oxides following deposition in rat lung. Environ. Res., 36: 359-378.

ROMNEY, E.M. & CHILDRESS, J.D. (1965) Effects of beryllium in plants and soils. Soil Sci., 100: 210-217.

ROMNEY, E.M., CHILDRESS, J.D., & ALEXANDER, G.V. (1962) Beryllium and the growth of bush beans. Science, 135: 786-787.

ROMNEY, E.M., WALLACE, A., ALEXANDER, G.V., & LUNT, O.R. (1980) Effect of beryllium on mineral element composition of bush beans. J. Plant Nutr., 2: 103-106.

ROSENKRANZ, H.S. & POIRIER, L.A. (1979) Evaluation of the mutagenicity and DNA-modifying activity of carcinogens and noncarcinogens in microbial systems. J. Natl Cancer Inst., 62: 873-892.

ROSKILL INFORMATION SERVICE (1980) The economics of beryllium, London, Roskill Information Services Ltd, 116 pp.

ROSS, W.D. & SIEVERS, R.E. (1972) Environmental air analysis for ultratrace concentrations of beryllium by gas chromatography. Environ. Sci. Technol, 6: 155-160.

ROSS, W.D., PYLE, J.L., & SIEVERS, R.E. (1977) Analysis for beryllium in ambient air particulates by gas chromatography. Environ. Sci Technol., 11: 467-471.

ROSSMAN, M.D., KERN, J.A., ELIAS, J.A., CULLEN, M.R., EPSTEIN, P.E., PREUSS, O.P., MARKHAM, T.N., & DANIELE, R.P. (1988) Proliferative response of bronchoalveolar lymphocytes to beryllium. A test for chronic beryllium disease. Ann. intern. Med., 108: 687-693.

SAINSBURY, C.L., HAMILTON, J.C., & HUFFMANN, C. (1968) Geochemical cycle of selected trace elements in the tin-tungsten-beryllium district, Western Seward Peninsula, Alaska - A reconnaissance study. Washington, DC, US Government Printing Office, 71 pp (US Geological Survey Bulletin 1242-F).

SALTINI, C., WINESTOCK, K., KIRBY, M., PINKSTON, P., & CRYSTAL, R.G. (1989) Maintainance of alveolitis in patients with chronic beryllium disease by beryllium-specific helper T cells. N. Engl. J. Med., 320: 1103-1109.

SANDERS, C.L., CANNON, W.C., POWERS, G.J., ADU, R.R., & MEIER, D.M. (1975) Toxicology of high-fired beryllium oxide inhaled by rodents. I. Metabolism and early effects. Arch. environ. Health, 30: 546-551.

SARACCI, R. (1985) Beryllium. In: Wald, N.J. & Doll, R., ed. Interpretation of negative epidemiological evidence for carcinogenicity. Lyon, International Agency for Research on Cancer, pp. 203-219 (IARC Scientific Publications No. 65).

SAUER, C. & LIESER, K.H. (1986) [Determination of trace elements in a raw water and in a drinking water.] Vom Wasser, 66: 277-284 (in German).

SCHEPERS, G.W.H. (1961) Neoplasia experimentally induced by beryllium compounds. Progr. exp. Tumor Res., 2: 203-244.

SCHEPERS, G.W.H. (1964) Biological action of beryllium. Reaction of the monkey to inhaled aerosols. Ind. Med. Surg., 33: 1-16.

SCHEPERS, G.W.H., DURKAN, T.M., DELAHANT, A.B., & CREEDON, F.T. (1957) The biological action of inhaled beryllium sulfate. Arch. ind. Health, 15: 32-58.

SCHONHERR, S. & PEVNY, I. (1985) [Beryllium allergy.] Arbeitsmed. Sozialmed. Präventivmed., 20: 281-286 (in German).

SCHORMULLER, J. & STAN, H.J. (1965) [Inhibition of various enzymes by beryllium salts.] Nahrung, 9: 435-444 (in German).

SCHRAMEL, P. & LI-QIANG, X. (1982) Determination of beryllium in the parts-per-billion range in three standard reference materials by inductively coupled plasma atomic emission spectrometry. Anal. Chem., 54: 1333-1336.

SCHROEDER, H.A. & MITCHENER, M. (1975a) Life-term studies in rats - effects of aluminium, barium, beryllium, and tungsten. J. Nutr., 105: 421-427.

SCHROEDER, H.A. & MITCHENER, M. (1975b) Life-term effects of mercury, methyl mercury, and nine other trace metals on mice. J. Nutr., 105(4): 452-458.

SCHUBERT, J. (1958) Beryllium and berylliosis. Sci. Am., **199**: 27-33.

SCOTT, D.R., LOSEKE, W.A., HOLBOKE, L.E., & THOMPSON, R.J. (1976) Analysis of atmospheric particulates for trace elements by optical emission spectrometry. Appl. Spectrosc., **30**: 392-405.

SCOTT, J.K. (1948) Pathologic anatomy of acute experimental beryllium poisoning. Arch. Pathol., **45**: 354.

SCOTT, J.K., NEUMANN, W.F., & ALLEN, R. (1950) The effect of added carrier on the distribution and excretion of soluble beryllium. J. biol. Chem., **182**: 291-298.

SENDELBACH, L.E., WITSCHI, H.P., & TRYKA, A.F. (1986) Acute pulmonary toxicity of beryllium sulfate inhalation in rats and mice: cell kinetics and histopathology. Toxicol. appl. Pharmacol., **85**: 248-256.

SHACKLETTE, H.T., HAMILTON, J.G., BOERNGEN, J.G., & BOWLES, J.M. (1971) Elemental composition of surficial materials in the conterminous United States. Washington, DC, US Government Printing Office, 71 pp (US Geological Survey, Professional Paper 574-D).

SIEM, P. (1886) [On the effects of beryllium and aluminium in animals.] Inaugural Dissertation, Dorpat (in German).

SIMMON, V.F. (1979a) *In vitro* mutagenicity assays of chemical carcinogens and related compounds with *Salmonella typhimurium*. J. Natl Cancer Inst., **62**: 893-899.

SIMMON, V.F. (1979b) *In vitro* assays for recombinogenic activity of chemical carcinogens and related compounds with *Saccharomyces cerevisiae* D3. J. Natl Cancer Inst., **62**: 901-909.

SIMMON, V.F., ROSENKRANZ, H.S., ZEIGER, E., & POIRIER, L.A. (1979) Mutagenic activity of chemical carcinogens and related compounds in the intraperitoneal host-mediated assay. J. Natl Cancer Inst., **62**: 911-918.

SIROVER, M.A. & LOEB, L.A. (1976) Metal-induced infidelity during DNA synthesis. Proc. Natl Acad. Sci. (USA) **73**(7): 2331-2335.

SKILLETER, D.N. (1984) Biochemical properties of beryllium potentially relevant to its carcinogenicity. Toxicol. Environ. Chem., **7**: 213-228

SKILLETER, D.N. (1986) Selective cellular and molecular effects of beryllium on lymphocytes. Toxicol. environ. Chem., **11**: 301-312.

SKILLETER, D.N. (1987) Beryllium. In: Fishbein, L., Furst, A., & Mehlman, M.A., ed. Genotoxic and carcinogenic metals: environmental and occupational occurrence and exposure. Princeton, New Jersey, Princeton Scientific Publishing Co., pp. 61-86 (Advances in Modern Environmental Toxicology, Vol. 11).

SKILLETER, D.N. & PRICE, R.J. (1984) Lymphocyte beryllium binding: Relationship to development of delayed beryllium hypersensitivity. Int. Arch. Allergy appl. Immunol., **73**(2): 181-183.

SLONIM, A.R. & RAY, E.E. (1975) Acute toxicity of beryllium sulfate to salamander larvae (*Ambystoma* spp.). Bull. environ. Contam. Toxicol., **13**: 307-312.

SLONIM, C.B. & SLONIM, A.R. (1973) Effect of water hardness on the tolerance of the guppy to beryllium sulfate. Bull. environ. Contam. Toxicol., **10**: 295-301.

SPENCER, H.C., JONES, J.C., SADEK, S.E., DODSON, K.B., & MORGAN, A.H. (1965) Toxicological studies on beryllium oxides. Toxicol. appl. Pharmacol., **7**: 498.

SPENCER, H.C., HOCK, R.H., BLUMENSHINE, J.V., MCCOLLISTER, S.B., SADEK, S.E., & JONES, J.C. (1968) Toxicological studies on beryllium oxides and beryllium containing exhaust products. Fairborn, Ohio, Wright Patterson Air Force Base, Aerospace Medical Research Laboratory (AMRL-TR-68-148).

SPENCER, H.C., MCCOLLISTER, S.B., KOCIBA, R.J., & HUMISTON, C.G. (1972) Toxicological studies on a beryllium-containing exhaust product. In: Proceedings of the 3rd Annual Conference on Environmental Toxicology, Ohio, Wright-Patterson Air Force Base, Aerospace Medical Research Laboratory, pp. 303-317 (AMRL-TR-72-130).

SPRINCE, N.L., KAZEMI, H., & HARDY, H.L. (1976) Current (1975) problem of differentiating between beryllium disease and sarcoidosis. Ann. N.Y. Acad. Sci., **278**: 654-664.

SPRINCE, N.L., KANAREK, D.J., & WEBER, A.L. (1978) Reversible respiratory disease in beryllium workers. Am. Rev. resp. Dis., **117**: 1011-1017.

STERNER, J.H. & EISENBUD, M. (1951) Epidemiology of beryllium intoxication. Arch. ind. Hyg., **4**: 123-151.

STIEFEL, T., SCHULZE, K., TOLG, G., & ZORN, H. (1976) [A combined method for the determination of beryllium in biological matrices by flameless atomic adsorption spectrometry.] Anal. Chim. Acta, **87**: 67-78 (in German).

STIEFEL, T., SCHULZE, K., ZORN, H., & TOLG, G. (1980a) Toxicokinetic and toxicodynamic studies of beryllium. Arch. Toxicol., **45**: 81-92.

STIEFEL, T., SCHULZE, K., TOLG, G., & ZORN, H. (1980b) Analysis of trace elements distributed in blood. Fresenius Z. anal. Chem., **300**: 189-196.

STOECKLE, J.D., HARDY, H.L., & WEBER, A.L. (1969) Chronic beryllium disease. Long- term follow-up of sixty cases and selective review of the literature. Am. J. Med., **46**: 545-561.

STOKINGER, H.E. (1981) The Metals: 5. Beryllium. In: Clayton, G.D. & Clayton, F.E., ed. Patty's industrial hygiene and toxicology, New York, John Wiley & Sons, Vol. 2A, pp. 1493-2060.

STOKINGER, H.E., SPRAGUE, G.F., HALL, R.H., ASHENBURG, N.J., SCOTT, J.K., & STEADMAN, L.T. (1950a) Acute inhalation toxicity of beryllium. I. Four definitive studies of beryllium sulfate at exposure concentrations of 100, 50, 10 and 1 mg per cubic meter. Arch. ind. Hyg. occup. Med., **1**: 379-397.

STOKINGER, H.E., ASHENBURG, N.J., DE VOLDRE, J., SCOTT, J.K., & SMITH, F.A. (1950b) Acute inhalation toxicity of beryllium. II. The enhancing effect of the inhalation of hydrogen fluoride vapor on beryllium sulfate poisoning in animals. Arch. ind. Hyg. occup. Med., **1**: 398-410.

STOKINGER, H.E., SPIEGL, C.J., ROOT, R.E., HALL, R.H., STEADMAN, L.T., STROUD, C.A., SCOTT, J.K., SMITH, E.A., & GARDNER, D.E. (1953) Acute inhalation toxicity of beryllium. IV: Beryllium fluoride at exposure concentrations of one and ten milligrams per cubic meter. Arch. ind. Hyg. occup. Med., 8: 493-506.

SUSSMAN, V.H., LIEBEN, J., & CLELAND, J.G. (1959) An air pollution study of a community surrounding a beryllium plant. Am. Ind. Hyg. Assoc. J., 20: 504-508.

SUTTON, W.R. (1939) Some changes produced in growth, reproduction, blood and urine of rats by salts of zinc with certain observations on the effects of cadmium and beryllium salts. Iowa State Coll. J. Sci., 14: 89.

TALLURI, M.V. & GUIGGIANI, V. (1967) Action of beryllium ions on primary cultures of swine cells. Caryologia, 20: 355-367.

TAPP, E. (1969) Beryllium-induced sarcomas of the rabbit tibia. Br. J. Cancer, 20: 778-783.

TARZWELL, C.M. & HENDERSON, C. (1960) Toxicity of less common metals to fishes. Ind. Wastes, 5: 12.

TAYLOR, M.L. & ARNOLD, E.L. (1971) Ultratrace determination of metals in biological specimens: quantitative determination of beryllium by gas chromatography. Anal. Chem., 43(10): 1328-1331.

TEPPER, L.B. (1972) Beryllium. CRC crit. Rev. Toxicol., 2: 235-259.

TEPPER, L.B., HARDY, H.L., & CHAMBERLIN, R.I. (1961) Toxicity of beryllium compounds, Amsterdam, Oxford, New York, Elsevier Science Publishers.

THOMAS, M. & ALDRIDGE, W.N. (1966) The inhibition of enzymes by beryllium. Biochem. J, 98: 94-99.

TOLLE, D.A., ARTHUR, M.F., & VAN VORIS, P. (1983) Microcosm/field comparison of trace element uptake in crops grown in fly ash-amended soil. Sci. total Environ., 31(3): 243-261.

TRUHAUT, R., FESTY, B., & LE TALAER, J.Y. (1968) Modalités de l'interaction du béryllium avec l'acide désoxiribonucléique et son incidence sur quelques systèmes enzymatiques. C.R. Acad. Sci. Sér. D, 266: 1192-1195.

TSUJII, H. & HOSHISHIMA, K. (1979) The effect of the administration of trace amounts of metals to pregnant mice upon the behaviour and learning of their offspring. Shinshu Daigaku Nogakubu Kiyo, 16: 13-27.

TURK, J.L. & POLAK, L. (1969) Experimental studies on metal dermatitis in guinea pigs. Int. Arch. Allergy, 36: 76-81.

US BUREAU OF MINES (1982) Minerals yearbook, 1981. Vol. I. Metals and minerals: beryllium. Washington, DC, US Department of the Interior, pp. 135-138.

US BUREAU OF MINES (1985a) Minerals yearbook, 1984, Vol. I. Metals and minerals: beryllium. Washington, DC, US Department of the Interior, pp. 153-157.

US BUREAU OF MINES (1985b) Beryllium - a chapter from mineral facts and problems. 1985 ed. Washington, DC, US Department of the Interior, 8 pp.

US BUREAU OF MINES (1986) Mineral commodity summaries 1986. Washington, DC, US Department of the Interior, pp. 20-21.

US EPA (1971) National inventory of sources and emissions: beryllium-1968,.Washington, DC, US Environmental Protection Agency, 56 pp (Publication No. APTD-1508).

US EPA (1973) Control techniques for beryllium air pollutants. Research Triangle Park, North Carolina, US Environmental Protection Agency, Office of Air and Water Programs, 75 pp (Report No. AP 116).

US EPA (1978a) National emission standard for beryllium. In: US code of Federal Regulations. Washington DC, US Government Printing Office, Title 40, Parts 61.30-61.32, pp. 276-277.

US EPA (1978b) In-depth studies on health and environmental impacts of selected water pollutants. Washington, DC, US Environmental Protection Agency (Contract No. 68-01-4646).

US EPA (1980) Ambient water quality criteria for beryllium. Washington, DC, US Environmental Protection Agency, Division of Water Planning and Standards (EPA Report 440/5-80-024).

US EPA (1987) Health assessment document for beryllium. Research Triangle Park, North Carolina, US Environmental Protection Agency, Office of Research and Development (EPA Report No. 600/8-84-026F).

VACHER, J. (1972) Immunological responses of guinea pigs to beryllium salts. J. med. Microbiol., 5(1): 91-108.

VAN CLEAVE, C.D. & KAYLOR, C.T. (1955) Distribution retention, and elimination of Be in the rat after intratracheal injection. Arch. ind. Health, 11: 375-392.

VAN ORDSTRAND, H.S., HUGHES, R., DE NARDI, J.M., & CARMODY, M.G. (1945) Beryllium poisoning. J. Am. Med. Assoc., 129: 1084-1090.

VASILYEVA, E.V., NIKITINA, L.S., & ORLOVA, A.A. (1977) Concentration of immunoglobulins in berylliosis. J. Hyg. Epidemiol. Microbiol. Immunol., 21: 254-260.

VENUGOPAL, B. & LUCKEY, T.D. (1978) Metal toxicity in mammals. Volume 2: Chemical toxicity of metals and metalloids. New York, London, Plenum Press, pp. 43-50.

VOISIN, G.A., COLLET, A., MARTIN, J.C., DANIEL-MOUSSARD, H., & TOULLET, F. (1964) Propriétés immunologiques de silice et de composés du béryllium: les formes solubles comparées aux formes insolubles. Rév. fr. Etud. clin. biol., 9: 819-828.

VORWALD, A.J., ed. (1950) Animal methods. In: Pneumoconiosis - beryllium, bauxite fumes, compensation. New York, Hoeber, pp. 393-425.

VORWALD, A.J. (1953) Adenocarcinoma in the lung of albino rats exposed to compounds of beryllium. In: Cancer of the lung, an evaluation of the problem. Proceedings of the Scientific Session, Annual Meeting of the American Cancer

Society, New York, November 1953. New York, American Cancer Society Inc., pp. 103-109.

VORWALD, A.J. (1962) Progress Report of the American Cancer Society. (unpublished report) (Grant No. E-253-C).

VORWALD, A.J. (1968) Biologic manifestations of toxic inhalants in monkeys. In: Vagtberg, H., ed. Use of nonhuman primates in drug evaluation. Austin, Texas, University of Texas Press, pp. 222-228.

VORWALD, A.J. & REEVES, A.L. (1959) Pathologic changes induced by beryllium compounds. Arch. ind. Health, 19: 190-199.

VORWALD, A.J., PRATT, P.C., & URBAN, E.C.J. (1955) The production of pulmonary cancer in albino rats exposed by inhalation to an aerosol of beryllium sulfate (Abstract). Acta Union Int. Contra Cancrum, 11: 735.

VORWALD, A.J., REEVES, A.L., & URBAN, E.C.J. (1966) Experimental beryllium toxicology. In: Stokinger, H.E., ed. Beryllium, its industrial hygiene aspects. New York, London, Academic Press, pp. 201-234.

WAGNER, W.D., GROTH, D.H., HOLTZ, J.L., MADDEN, G.E., & STOKINGER, H.E. (1969) Comparative chronic inhalation toxicity of beryllium ores, bertrandite, and beryl, with production of pulmonary tumors by beryl. Toxicol. appl. Pharmacol., 15: 10-29.

WAGONER, J.K., INFANTE, P.F., & BAYLISS, D.L. (1980) Beryllium: An etiologic agent in the induction of lung cancer, nonneoplastic respiratory disease, and heart disease among industrially exposed workers. Environ. Res., 21: 15-34.

WALSH, K. & REES, G.H. (1978) Beryllium compounds. In: Kirk, R.E. & Othmer, D.F., ed. Encyclopedia of chemical technology. 3rd ed. New York, Chichester, John Wiley & Sons, Vol. 3, pp. 824-829.

WEBER, H. & ENGELHARDT, W.E. (1933) [On a dust-generating device of high accurancy and a method for the microgravimetric determination of dust. Application for the investigation of dusts from beryllium production.] Gewerbehyg. Unfallverhüt., 10: 41-47 (in German).

WEDEPOHL, K.H., ed. (1966) Handbook of geochemistry. Vol. II-I, Section 4: Beryllium. New York, Heidelberg, Berlin, Springer-Verlag.

WHITE, M.R., FINKEL, A.J., & SCHUBERT, J. (1951) Protection against experimental beryllium poisoning by aurintricarboxylic acid. J. Pharmacol. exp. Ther., 102: 88-93.

WHO (1984) Guidelines for drinking-water quality. Vol. 2. Health criteria and other supporting information. Geneva, World Health Organization, pp. 80-83.

WHO (1985) Environmental health criteria 51: Guide to short-term tests for detecting mutagenic and carcinogenic chemicals. 208 pp., Geneva, World Health Organization.

WHO (1990) Health and safety guide no. 44: Beryllium. Geneva, World Health Organization.

WICKS, S.A. & BURKE, R.W. (1977) Determination of beryllium by fluorescence spectrometry. In: Mavrodineanu, R., ed. Procedures used at the National Bureau of Standards to determine selected trace elements in biological and botanical materials. Washington, DC, US Department of Commerce, pp. 85-89 (NBS Special Publication No. 492).

WILKE, B.-M. (1987) [Effects of inorganic pollutants on microbial processes], Berlin (West), Umweltbundesamt (Research Report No. 10701006) (in German).

WILLIAMS, D., JONES WILLIAMS, W., & WILLIAMS, J.E. (1969) Enzyme histochemistry of epitheloid cells in sarcoidosis and sarcoid like granulomas. J. Pathol., **97**: 705-709.

WILLIAMS, G.M., LASPIA, M.F., & DUNKEL, V.C. (1982) Reliability of the hepatocyte primary culture/DNA repair test in testing of coded carcinogens and noncarcinogens. Mutat. Res., **97**: 359-370.

WILLIAMS, R.J.B. & LE RICHE, H.H. (1968) The effect of traces of beryllium on the growth of kale, grass and mustard. Plant Soil, **29**: 317-326.

WILLIAMS, W.R. & JONES WILLIAMS, W. (1982) Comparison of lymphocyte transformation and macrophage migration inhibition tests in the detection of beryllium hypersensitivity. J. clin. Pathol., **35**: 684-687.

WINDHOLZ, M., ed. (1976) The Merck index. 9th ed. Rahway, New Jersey, Merck & Co., pp. 153-155.

WITSCHI, H.P. (1970) Effects of beryllium on deoxyribonucleic acid-synthesizing enzymes in regenerating rat liver. Biochem. J., **120**: 623-634.

WITSCHI, H.P. & ALDRIDGE, W.N. (1968) Uptake, distribution, and binding of beryllium to organelles of the rat liver cell. Biochem. J., **106**: 811-820.

WITSCHI, H.P. & MARCHAND, P. (1971) Interference of beryllium with enzyme induction in rat liver. Toxicol. appl. Pharmacol., **20**(4): 565-572.

WOHLER, F. (1828) [On beryllium and yttrium.] Pogg. Ann., **13**: 577-582 (in German).

WOLF, W.R., TAYLOR, M.I., HUGHES, B.M., TIERNAN, T.O., & SIEVERS, R.E. (1972) Determination of chromium and beryllium at the picogram level by gas chromatography-mass spectrometry. Anal. Chem., **44**: 616-618.

WOLNIK, K.A., FRICKE, F.L., & GASTON, C.M. (1984) Quality assurance in the elemental analysis of foods by inductively coupled plasma spectroscopy. Spectrochim. Acta Part B, **398**: 649-655.

WOOD, J.M. & WANG, H.K. (1983) Microbial resistance to heavy metals. Environ. Sci. Technol., **17**: 582A-590A.

WURM, H. & RUGER, H. (1942) [Experiments concerning the question of beryllium dust pneumonia.] Beitr. klin. Tuberk., **98**: 396-404 (in German).

YAMAGUCHI, S. (1963) Studies on osteogenic sarcoma induced experimentally by beryllium. Nagasaki Igakkai Zasshi, **38**: 127-139.

YAVOROVSKAYA, S.F. & GRINBERG, K.M. (1974) [Detecting aerosol metals (beryllium, chromium, aluminium) in the air by gas chromatography.] Gig. i Sanit., **11**: 54-57 (in Russian, with English summary).

ZAKOUR, R.A., TKESHVILI, L.K., SHERMAN, C.W., KOPLITZ, R.M., & LOEB, L.A. (1981) Metal-induced infidelity of DNA synthesis. J. cancer Res. clin. Oncol., **99**: 187-196.

ZDROJEWSKI, A., DUBOIS, L., & QUICKERT, N. (1976) Reference method for the determination of beryllium in airborne particulates. Sci. Total Environ., **6**: 165-173.

ZIELINSKI, J.F. (1961) A summary of the results of seven years of experience in investigating the dispersion of beryllium in the air of a modern alloy foundry. Workshop on beryllium, 1961. Cincinnati, Ohio, The Kettering Laboratory, Univ. of Cincinnati, pp. 84-102.

ZORN, H. & DIEM, H. (1974) [Importance of beryllium and its compounds in occupational medicine.] Zentralbl. Arbeitsmed. Arbeitsschutz, **24**: 3-8 (in German)

ZORN, H., STIEFEL, T., & DIEM, H. (1977) [Importance of beryllium and its compounds in occupational medicine, 2nd report.] Zbl. Arbeitsmed., **27**: 83-88 (in German).

Résumé

RESUME ET CONCLUSIONS

1. Identité, propriétés physiques et chimiques, méthodes d'analyse

Le béryllium est un métal gris acier, fragile, qui n'existe à l'état naturel que sous la forme d'un seul isotope, le ^{9}Be. Ses composés sont bivalents. C'est un élément unique par certaines de ses propriétés. Ainsi il est le plus léger de tous les corps solides et chimiquement stables, avec un point de fusion, une chaleur spécifique, une chaleur de fusion, une charge de rupture exceptionnellement élevés. Il possède une excellente conductivité électrique et thermique. Du fait de son faible numéro atomique, le béryllium est très perméable aux rayons X. Parmi ses propriétés nucléaires on peut citer la rupture, la diffusion et la réflection de neutrons ainsi que l'émission de neutrons par bombardement alpha.

Le béryllium partage un certain nombre de propriétés chimiques avec l'aluminium, en particulier sa forte affinité pour l'oxygène. Par suite, il se forme à la surface du métal et de ses alliages, une péllicule très stable d'oxyde de béryllium (BeO), qui leur confère une très grande résistance à la corrosion, à l'eau et aux acides oxydants à froid. A l'état pulvérulent, le béryllium brûle dans l'oxygène à une température de 4500 °C. L'oxyde de béryllium fritté est très stable et possède les propriétés d'une céramique. Les sels dans lesquels il se trouve à l'état cationique s'hydrolisent dans l'eau et réagissent pour former des hydroxydes insolubles ou des complexes hydratés, aux valeurs du pH comprises entre 5 et 8 et des béryllates au-dessus de pH 8.

Aux alliages, le béryllium apporte un ensemble de propriétés remarquables, en particulier: résistance à la corrosion, module élevé d'élasticité, amagnétisme, propriétés antiétincelantes, forte conductivité électrique et thermique et meilleure résistance à la rupture que l'acier.

On utilise diverses méthodes d'analyse pour la recherche et le dosage du béryllium dans les différents milieux. Parmi les méthodes anciennes on peut citer la spectroscopie, la fluorimétrie et la spectrophotométrie. Les méthodes de choix sont la spectrométrie d'absorption atomique sans flamme ainsi que la chromatographie en phase gazeuse; les limites de détection sont de 0,5 ng/échantillon (absorption atomique sans flamme) et de 0,04 pg/échantillon (chromatographie en phase gazeuse avec détection par capture d'électrons). En outre, la spectrométrie d'émission atomique à plasma avec couplage par induction s'utilise de plus en plus.

2. Sources d'exposition humaine et environnementale

Le béryllium se situe au trente-cinquième rang des éléments par ordre d'abondance dans la croûte terrestre avec une teneur moyenne d'environ 6 mg/kg. Exception faite des pierres précieuses, de l'émeraude (béryl contenant du chrome) et de l'aigue-marine (béryl contenant du fer), il n'existe que deux minéraux qui présentent une importance économique. Le béryl contient jusqu'à 4% de béryllium et il est extrait en Argentine, au Brésil, en Chine, en Inde, au Portugal, en URSS et dans plusieurs pays d'Afrique australe et centrale. La bertrandite, dont la teneur en béryllium est inférieure à 1%, est cependant devenue la principale source de ce métal aux Etats-Unis d'Amérique.

La production annuelle mondiale de minerais de béryllium au cours de la période 1980–84 a été évaluée à environ 10000 tonnes, ce qui correspond approximativement à 400 tonnes de béryllium. Malgré les fluctuations considérables de l'offre et de la demande de béryllium qui résultent de programmes sporadiques de la part des gouvernements dans le domaine des armements, de l'énergie nucléaire et des activités aérospatiales, on a estimé, en 1986, que la demande de béryllium allait vraisemblablement augmenter de 4% par an en moyenne jusqu'en 1990.

D'une façon générale, les émissions de béryllium au cours de la production et de l'utilisation de ce métal sont peu importantes par rapport à celles qui se produisent lors de la combustion du charbon et du mazout, qui contiennent respectivement 1,8 à 2,2 mg de béryllium/kg de poids à sec et jusqu'à 100 µg de béryllium/litre. Les émissions de béryllium résultant de la combustion des combustibles fossiles correspond aux Etats-Unis, l'un des principaux pays producteurs, à environ 93% de l'ensemble des émissions de

Résumé

béryllium. Des mesures de contrôle plus efficaces peuvent réduire notablement les émissions de béryllium par les centrales thermiques.

La concentration de fond en béryllium dans l'air ambiant dépend essentiellement de la combustion des combustibles fossiles, mais les émissions provenant d'unités de production peuvent conduire localement à des concentrations élevées en particulier lorsque les mesures antipollution sont insuffisantes. De même, des émissions locales non négligeables peuvent se produire lors de l'essai et de l'utilisation de fusées utilisant des propergols à base de béryllium. Dans l'industrie, l'exposition se produit principalement lors du traitement des minerais de béryllium, du béryllium métallique, des alliages à base de béryllium et de l'oxyde de béryllium. Le béryllium n'est produit qu'au Japon, aux Etats-Unis et en URSS. D'autres pays importent le métal, son oxyde ou des alliages aux fins de transformations ultérieures.

La plupart des déchets de béryllium font l'objet de mesures de lutte antipollution et sont recyclés ou enfouis dans le sol. La plupart des produits de transformation ne peuvent pas être recyclés car leur faible volume et leur basse teneur en béryllium font que cette opération n'est pas rentable.

Environ 72% de la production mondiale de béryllium est utilisée sous la forme de cupro-béryllium et d'autres alliages dans les industries aérospatiales, électroniques et mécaniques. Le reste est utilisé sous forme d'oxyde de béryllium pour la fabrication de céramiques utilisées principalement en électronique et en microélectronique.

3. **Transport, distribution et transformation dans l'environnement**

On ne dispose que de données limitées sur la destinée du béryllium dans l'environnement. Les particules d'oxyde de béryllium en suspension dans l'atmosphère retombent au sol soit à sec, soit avec les précipitations. Dans l'environnement, où le pH varie de 4 à 8, le béryllium est fortement absorbé par les minéraux sédimentaires finement dispersés, ce qui en empêche le passage dans les eaux souterraines.

On ne pense pas que le béryllium subisse une bioamplification par l'intermédiaire de la chaîne alimentaire. La plupart des plantes fixent le béryllium présent dans le sol en petites quantités et la proportion qui parvient jusqu'aux racines ou en d'autres parties de la plante reste extrêmement faible.

4. Niveaux dans l'environnement et exposition humaine

La concentration du béryllium dans les eaux de surface et les eaux de consommation est généralement de l'ordre du µg/litre. Dans le sol, ces concentrationns se situent entre 1 et 7 mg/kg. Les plantes terrestres en contiennent généralement moins de 1mg/kg de poids à sec. On a trouvé dans divers organismes marins des quantités allant jusqu'à 100 µg/kg de poids frais.

En zone rurale aux Etats-Unis on a relevé des concentrations atmosphériques allant de 0,03 à 0,06 ng/m^3. Dans les pays où on utilise moins de combustibles fossiles, les concentrations de fond devraient être plus faibles. Aux Etats-Unis, les concentrations moyennes annuelles dans l'air des villes vont de moins de 0,1 à 6,7 ng/m^3. Dans des villes japonaises, on a trouvé une moyenne de 0,04 ng/m^3 avec des valeurs maximales (0,2 ng/m^3) dans les zones industrielles.

Avant la mise en place de mesures antipollution dans les années 1950, les concentrations atmosphériques de béryllium étaient très fortes au voisinage des unités de production et de transformation de ce métal. En outre, on notait une exposition "para-professionnelle" dans les familles des travailleurs que l'on qualifiait de "cas de voisinage" et qui était due soit à un contact avec les vêtements des travailleurs, soit à une exposition atmosphérique, soit aux deux. Aujourd'hui, ce type d'exposition est négligeable pour la population dans son ensemble. La principale source d'exposition environnementale impliquant la population dans son ensemble est due à la présence de béryllium atmosphérique résultant de l'utilisation de combustibles fossiles. Il peut se produire des expositions exceptionnellement élevées au voisinage de centrales thermiques où l'on brûle du charbon à forte teneur en béryllium et où les mesures antipollution ne sont pas suffisamment strictes. La fumée de tabac constitue probablement aussi une importante source d'exposition au béryllium.

Résumé

L'utilisation croissante de béryllium en art dentaire pour la confection d'appareils pourrait également jouer un certain rôle dans l'exposition de la population générale du fait que le béryllium a une forte tendance à provoquer des allergies de contact.

Avant 1950, l'exposition professionnelle au béryllium était généralement très importante et il n'était pas rare de rencontrer des concentrations dépassant 1 mg/m^3. Les mesures mises en place dans divers pays pour satisfaire aux normes d'exposition professionnelle fixées à 1–5 µg de Be/m^3 (en moyenne pondérée par rapport au temps) ont considérablement réduit la concentration du béryllium sur les lieux de travail, encore que ces valeurs ne soient pas partout respectées.

La teneur des tissus ou des liquides biologiques en béryllium peut être l'indice d'une exposition antérieure. Chez des personnes qui ne sont pas particulièrement exposées, les concentrations urinaires dont d'environ 1 µg/litre et les concentrations dans le tissu pulmonaire de moins de 20 µg/kg (de poids à sec). Les quelques données dont on dispose permettent pas d'établir une relation bien nette entre l'exposition et la charge de l'organisme, encore que l'on trouve indiscutablement des teneurs élevées (supérieures à 20 µg/kg) dans les tissus pulmonaires de patients atteints de bérylliose.

5. Cinetique et metabolisme

On ne dispose d'aucune donnée sur le dépot et l'absorption du béryllium après inhalation. Les études sur les animaux de laboratoire ont montré qu'après dépot dans les poumons, le béryllium demeure en place et passe peu à peu dans le sang. L'élimination pulmonaire est biphasique, avec une phase d'élimination rapide au cours de la première et de la deuxième semaine suivant l'arrêt de l'exposition.

La majeure partie du béryllium circulant dans le courant sanguin s'y trouve sous forme de phosphate colloïdal. Une part non négligeable de la dose inhalée est incorporée au squelette qui constitue la site final d'accumulation du béryllium. En général, l'exposition par voie respiratoire entraîne également une accumulation prolongée de quantités importantes de béryllium dans le tissu pulmonaire, plus particulièrement dans les ganglions lymphatiques. Les dérivés plus solubles vont également se fixer dans d'autres tissus ou organes: foie, ganglions lymphatiques abdominaux, rate, coeur, muscles, peau et reins. Après administration de béryllium par voie orale, une faible part de celui-ci (moins de 1% de la dose) passe en général dans le sang et aboutit au squelette. On en trouve également de petites quantités dans les voies digestives et le foie.

La résorption du béryllium par la peau intacte est négligeable car le béryllium se lie aux constituants de l'épiderme.

Une fois absorbé, le béryllium est en très grande partie rapidement éliminé dans les urines et en plus faible quantité dans les matières fécales. Le béryllium rejeté par la voie fécale provient probablement de l'ingestion des particules éliminées des voies respiratoires.

Du fait de la rétention du béryllium au niveau sequelettique et pulmonaire, sa demi-vie biologique est extrêmement longue. On estime par exemple que chez l'homme elle est de 450 jours dans le squelette.

6. Effets sur les etres vivant dans leur milieu naturel

Les microorganismes terricoles, cultivés dans un milieu pauvre en magnésium, se développent mieux en présence de béryllium en raison de la substitution partielle du béryllium au magnésium dans le métabolisme de ces microorganismes. Des effets stimulants de ce genre ont également été observés chez des algues et des plantes cultivées. Le phénomène semble dépendre du pH car il ne se produit qu'à pH élevé. A pH inférieur ou égal à 7, le béryllium est toxique pour les plantes aquatiques et terrestres, quelle que soit la teneur en magnésium du milieu de culture.

En général, la croissance végétale est inhibée par la présence de composés solubles du béryllium à des concentrations de l'ordre du mg/litre. Par exemple dans le cas du haricot commun (Phaseolus vulgaris) cultivé dans une solution nutritive à pH 5,3, on observe une réduction de 88% du rendement pour une concentration de 5 mg Be/litre. Les effets s'observent d'abord au niveau des racines qui virent au brun et ne reprennent pas leur croissance normale. C'est à ce niveau que la majeure partie du béryllium s'accumule, les quantités qui passent dans les parties supérieures de la plante étant très faibles. On estime à 3000 mg Be/kg au niveau des racines et à 6 mg Be/kg au niveau des feuilles extérieures du chou (*Brassica oleracea*) en poids à sec, la teneur en béryllium qui entraîne une diminution de 50% du rendement.

Un rabougrissement des racines et des feuilles a été noté dans des cultures de haricots, de blé et de trèfle ladino, mais sans chlorose ni tavelure des feuilles.

Dans les cultures en pleine terre, la phytotoxicité du béryllium dépend de la nature du sol, en particulier de sa capacité à échanger les cations et de son pH. A côté de l'effet de substitution au magnésium, la réduction de la phytotoxicité en milieu alcalin résulte également de la précipitation de béryllium sous forme de phosphate non utilisable.

Le mécanisme qui est à la base de la phytotoxicité du béryllium repose probablement sur l'inhibition de certaines enzymes spécifiques de la plante, en particulier des phosphatases. Le béryllium inhibe également la fixation de certains ions minéraux essentiels.

Les études de toxicité aiguë effectuées sur diverses espèces de poissons d'eau douce ont montré que la CL_{50} allait de 0,15 à 2 mg Be/litre selon l'espèce et les conditions expérimentales. La toxicité pour les poissons augmente en raison inverse de la dureté de l'eau; le sulfate de béryllium est dix à cent fois plus toxique pour les vairons et pour *Lepomis macrochirus* dans l'eau douce que dans l'eau dure. Les larves de salamandre et la daphnie (*Daphnia magna*) présentent une sensibilité analogue.

On ne dispose d'aucune donnée validée sur la toxicité chronique du béryllium pour les animaux aquatiques encore qu'une étude non publiée ait montré que la daphnie pouvait souffrir de concentrations beaucoup plus faibles de béryllium (5 µg Be/litre), lors d'études de reproduction à long terme que lors d'études de toxicité aiguë (CE_{50} = 2500 µg Be/litre).

7. **Effets sur les animaux d'experience et les systemes d'epreuves** *in vitro*

Chez l'animal d'expérience, les symptômes d'une intoxication aiguë par le béryllium se caractérisent par des troubles respiratoires, des spasmes, un choc hypoglycémique et une paralysie respiratoire.

L'implantation de dérivés du béryllium et de béryllium métallique dans les tissus sous cutanés peut produire des granulomes analogues à ceux que l'on observe chez l'homme. On a pu faire apparaître une hypersensibilité cutanée chez des cobayes par injection intradermique de composés solubles du béryllium.

Administré à des ratons, le carbonate de béryllium détermine indirectement un rachitisme chez ces animaux, en effet la précipitation intestinale du phosphate de béryllium entraîne une carence en phosphate.

On a observé chez diverses espèces animales l'apparition d'une pneumonie chimique aiguë après inhalation de béryllium métallique ou de divers composés du béryllium, notamment de dérivés insolubles. Une exposition quotidienne répétée à une nébulisation de sulfate de béryllium à la concentration moyenne de 2 mg/m^3 s'est révélée mortelle pour des rats (mortalité de 90%), des chiens (80%), des chats (80%), des lapins (10%), des cobayes (60%), des singes (100%), des chèvres (100%), des hamsters (50%) et des souris (10%). En raison de l'effet synergisant de l'ion fluorure, les effets du fluorure de béryllium ont été à peu près deux fois plus intense que ceux du sulfate. Certaines de lésions observées dans les poumons ressemblaient à celles qu'on voit chez l'homme mais les granulomes n'étaient pas identiques.

La toxicité de l'oxyde de béryllium insoluble par la voie respiratoire dépend en grande partie de ses propriétés physiques et chimiques, lesquelles peuvent beaucoup varier selon les conditions de production. Du fait de la granulométrie plus fine de l'oxyde de béryllium produit à basse température (400 °C), qui entraîne une moindre agrégation, une dose de 3,6 mg Be/m^3 pendant 40 jours a provoqué une certaine mortalité chez des rats et des lésions pulmonaires marquées chez des chiens. En revanche l'administration de deux qualités d'oxyde de béryllium produits à haute température (1350 °C et 1150 °C, respectivement) n'a pas produit de lésions pulmonaires malgré une exposition totale plus forte (32 mg Be/m^3 pendant 360 heures).

(1350 °C et 1150 °C, respectivement) n'a pas produit de lésions pulmonaires malgré une exposition totale plus forte (32 mg Be/m^3 pendant 360 heures).

La réaction non maligne caractéristique à une exposition prolongée par voie respiratoire à de faibles concentrations de dérivés du béryllium solubles ou insolubles, est une pneumonie chronique avec granulomes, qui ne correspond que partiellement à la maladie chronique observée chez l'homme.

Les épreuves de génotoxicité effectuées sur cultures de cellules somatiques de mammifères montrent que le béryllium réagit sur l'ADN et provoque de mutations géniques, des aberrations chromosomiques et des échanges entre chromatides soeurs; en revanche il n'est pas mutagène dans les systèmes d'épreuves bactériens.

L'injection de béryllium à des lapins par voie intraveineuse (3,7 à 700 mgBe) et intramédullaire (0,144–216 mg Be), soit sous forme métallique soit sous la forme de dérivés, a entraîné l'apparition d'ostéosarcomes et de chondrosarcomes avec, chez 40 à 100% des animaux, des métastases siégeant le plus souvent dans les poumons.

Chez des rats, l'inhalation (0,8–9000 μg Be/m^3) ou l'intubation trachéenne (0,3-9 mg Be) de béryllium métallique, de dérivés solubles et insolubles, et de divers alliages à base de béryllium, a provoqué l'apparition de tumeurs pulmonaires de type adénome ou adénocarcinome, dont certains donnaient lieu à des métastases. Le béryl (620 μg Be/m^3) a été au cours de cette étude le seul minerai de béryllium capable de provoquer l'apparition de cancers du poumon (ce n'était pas le cas en revanche de la bertrandite à 210 μg Be/m^3). L'oxyde de béryllium s'est révélé cancérogène pour le rat, mais l'incidence des adénocarcénomes pulmonaires était beaucoup plus forte après administration intratrachéenne (9 mg Be) d'une qualité d'oxyde produite à basse température (51%) qu'avec des oxides produits à haute température (11 à 16%). A l'époque où ces études ont été réalisées, elles n'étaient pas conçues ni menées selon les critères actuels et les données correspondantes doivent donc être considérées avec beaucoup de prudence.

L'induction de cancers pulmonaires par le béryllium est très spécifique d'espèce. Alors que les rats et éventuellement les singes sont très réceptifs à cet égard, on n'a pas observé de tumeurs du poumon chez les lapins, les hamsters ni les cobayes.

Trois théories ont été avancées pour expliquer la toxicité du béryllium: 1) le béryllium affecterait le métabolisme du phosphate en inhibant des enzymes clés, en particulier les phosphatases alcalines; 2) le béryllium inhiberait le réplication et la prolifération cellulaires en bloquant les enzymes du métabolisme des acides nucléiques; enfin 3) la toxicité du béryllium serait due à un mécanisme immunitaire, comme le montre l'apparition d'une hypersensibilité cutanée à médiation cellulaire chez le cobaye.

8. Effets sur l'homme

Seule l'exposition au béryllium sur les lieux de travail présente un intérêt toxicologique. Avant que ne soient prises dans les unités de production de béryllium des mesures de limitations des émissions et autres mesures d'hygiène, plusieurs cas de "bérylliose de voisinage" avaient été signalés. En 1966 on avait ainsi fait état de 60 cas aux Etats-Unis d'Amérique dont certains avaient pu être imputés à des contacts avec les vêtements de travailleurs affectés à ces unités de production (exposition paraprofessionnelle) ou à une exposition atmosphérique au voisinage des unités de production. Aucun cas de ce type n'a été signalé au cours des dernières années.

Récemment, on a signalé plusieurs cas de stomatites allergiques dus probablement à des prothèses dentaires à base de béryllium.

Au cours des années 1930 et 1940, plusieurs centaines de cas de bérylliose aiguë se sont déclarés, en particulier chez des travailleurs employés dans des unités d'extraction du béryllium en Allemagne, en Italie, aux Etats-Unis et en URSS. L'inhalation de sels solubles de béryllium, en particulier le fluorure et le sulfate à des concentrations supérieures à 100 $\mu g\, Be/m^3$, produisait systématiquement des symptômes aigus chez presque tous les travailleurs exposés, alors qu'aux concentrations inférieures ou égales à 15 $\mu g/m^3$ (déterminées avec des méthodes d'analyse aujourd'hui périmées) aucun cas n'était enregistré. Après l'adoption, au début des années 1950, d'une limite maximale d'exposition de 25 $\mu g/m^3$, on a constaté une diminution très marquée des cas de bérylliose aiguë.

Résumé

La symptomatologie le la bérylliose aiguë comporte des manifestations qui vont de la simple inflammation des muqueuses nasales et pharyngées à la trachéobronchite et à la pneumonie chimique grave. Dans les cas graves, les malades peuvent mourir de pneumonie aiguë mais la plupart du temps, la guérison est totale une à quatre semaines après cessation de l'exposition. Dans quelques rares cas, une bérylliose chronique peut se manifester plusieurs années après guérison de la forme aiguë.

Un contact direct avec des composés solubles entraîne une dermatite de contact et éventuellement une conjonctivite. Les individus sensibles réagissent beaucoup plus rapidement et à des concentrations plus faibles. Introduit dans ou sous l'épiderme, les composés solubles ou insolubles du béryllium produisent des ulcérations chroniques avec apparation fréquente de granulomes au bout de plusieurs années.

La bérylliose chronique se distingue de la forme aiguë par sa période de latence qui peut aller de quelques semaines à plus de 20 ans; elle est longue et de gravité progressive. Le US Beryllium Case Registry (Registre des cas de bérylliose des Etats-Unis d'Amérique) constitue un fichier central où sont enregistrés les cas de bérylliose. Il a été créé en 1952 et 888 cas y ont été consignés entre cette date et 1983. Parmi ces cas, 622 ont été classés comme chroniques, dont 557 attribuables à une exposition professionnelle, soit dans l'industrie des lampes à fluorescence (319 cas) soit dans des unités d'extraction du béryllium (101 cas). Après l'abandon en 1949 de l'utilisation de silicate double de zinc et de béryllium et d'oxyde de béryllium pour la production de phosphores destinés aux tubes à fluorescence et la fixation d'une limite d'exposition professionnelle (TWA = $2 \mu g\, Be/m^3$), les cas de bérylliose chronique ont diminué de façon spectaculaire, mais on a en revanche enregistré de nouveaux cas qui résultaient d'une exposition à du béryllium présent dans l'atmosphère à la concentration d'environ $2 \mu g/m^3$.

Il est préférable de parler d'ailleurs de "maladie chronique du béryllium" plutôt que de "bérylliose" car il s'agit d'une maladie différente d'une pneumoconiose typique. Les caractéristiques principales en sont une inflammation granulomateuse du poumon, associée à une dyspnée d'effort, à de la toux, des douleurs thoraciques, une fatigue et une faiblesse générales. On peut également observer une hypertrophie ventriculaire droite avec insuffisance cardiaque, une hépatomégalie, une splénomégalie, une cyanose et un hippocratisme digital. On a également constaté que la maladie s'accompagnait de modifications des protéines sériques et de la fonction hépatique, de calculs rénaux et d'ostéosclérose. L'évolution n'est pas uniforme; quelques fois il y a rémission spontanée pendant des semaines ou des années, puis exacerbation. Dans la majorité des cas, on observe une pneumopathie progressive avec risque accru de décès par insuffisance cardiaque ou respiratoire. Le taux de morbidité chez les ouvriers de l'industrie du béryllium varie de 0,3 à 7,5%. Chez les patients atteints de la maladie chronique, la mortalité peut monter jusqu'à 37%.

L'examen macroscopique des poumons révèle des altérations diffuses avec fibrose interstitielle et envahissement du parenchyme par de petits nodules disséminés. Du point du vue histologique, on observe de granulomes de type sarcoïde avec une inflammation interstitielle de degré variable; ces aspects ne se distinguent généralement pas de ceux qu'offrent les autres granulomatoses telles que la sarcoïdose ou la tuberculose.

L'anamnèse et l'analyse des tissus sont utiles au diagnostic de la bérylliose encore que la présence de béryllium dans les prélèvements ne soit pas une preuve de la maladie. La cuti-réaction n'est pas recommandée car il s'agit là d'une méthode peu fiable et très sensibilisante. Les examens de laboratoire les plus utilisés sont l'épreuve d'inhibition de la migration des macrophages et le test de transformation lymphoblastique.

Ces méthodes de mesure de l'hypersensibilité reposent sur un mécanisme immunitaire qui est probablement à la base de la bérylliose chronique et de l'hypersensibilité cutanée et granulomateuse retardées.

Les variations considérables du temps de latence et l'absence de relation dose-réponse dans la bérylliose chronique peuvent s'expliquer par une sensibilisation immunologique. Il semble que la grossesse joue le rôle d'un "facteur de stress" précipitant car 66% des 95 femmes figurant parmi les cas mortels de bérylliose enregistrés dans le US Beryllium Case Registry, étaient effectivement enceintes.

béryllium, l'industrie des céramiques (production et recherche), la recherche et la production d'énergie. Les normes actuelles d'exposition professionnelle ne protègent pas véritablement les individus sensibles contre la bérylliose chronique.

Lors d'un certain nombre d'études épidémiologiques, on a examiné la cancérogénicité du béryllium chez les ouvriers de deux unités de production aux Etats-Unis d'Amérique et l'on a également compulsé un registre de cas de bérylliose pulmonaire, où figuraient des employés de ces unités et d'autres branches d'activité. Les résultats de ces études ont été discutés en raison de la possibilité d'un biais sélectif, de l'existence de facteurs de confusion dus au tabagisme et de la sous estimation du nombre attendu de décès par cancer du poumon, étant donné que les taux de mortalité pour la période 1965–67 avaient été utilisés pour calculer la mortalité prévue en 1968–75. Il est peu probable que les deux premiers points aient pu contribuer de façon importante à accroître le risque de cancer du poumon; en revanche, les données qui figurent dans ce document sont fondées sur des prévisions "corrigées" du nombre de décès par cancer du poumon. Toutes les études effectuées ont fait ressortir une augmentation significative du risque de cancer du poumon.

9. Evaluation des risques pour la sante humaine et des effets sur l'environnement

9.1 Risques pour la sante humaine

Dans la mesure où l'industrie du béryllium applique des mesures antipollution convenables, l'exposition de la population générale se limite actuellement à de faibles concentrations de béryllium atmosphérique résultant de l'utilisation de combustibles fossiles. Dans certains cas exceptionnels, où l'on utilise un charbon excessivement riche en béryllium, il pourrait se poser des problèmes de santé. L'utilisation du béryllium pour la confection de prothèses dentaires est à revoir du fait du pouvoir sensibilisateur très important de cette substance.

Les cas de bérylliose aiguë se manifestant sous la forme de rhinopharyngites, de bronchites et de pneumonies chimiques graves se sont réduits de façon spectaculaire et aujourd'hui, ils ne pourraient

Les cas de bérylliose aiguë se manifestant sous la forme de rhinopharyngites, de bronchites et de pneumonies chimiques graves se sont réduits de façon spectaculaire et aujourd'hui, ils ne pourraient se produire qu'en cas de panne des systèmes antipollution. La bérylliose chronique se distingue de la forme aiguë par sa très longue période de latence qui peut aller de quelques semaines à plus de 20 ans et par sa gravité progressive. C'est essentiellement les poumons qui sont touchés. Cette affection se caractérise par une inflammation granulomateuse du tissu pulmonaire avec dyspnée d'effort, toux, douleurs thoraciques, perte de poids et faiblesse générale. Les effets sur les autres organes sont probablement dus à des causes indirectes. Cette affection peut encore s'observer chez des individus sensibilisés exposés à des concentrations d'environ 2 $\mu g/m^3$; elle se caractérise par de grandes variations dans le temps de latence et l'absence de relation dose-réponse.

Malgré un certain nombre d'insuffisances dans la conception des études et les pratiques de laboratoire, l'activité cancérogène du béryllium chez diverses espèces animales est confirmée.

Un certain nombre d'études épidémiologiques ont montré que l'exposition professionnelle au béryllium comportait un risque accru de cancer du poumon. L'interprétation des résultats obtenus a fait l'objet d'un certain nombre de critiques mais les données disponibles permettent de conclure que c'est très vraisemblablement le béryllium qui est à l'origine de l'accroissement du risque de cancer pulmonaire observé chez les travailleurs exposés.

9.2 Effets sur l'environnement

On ne dispose que de données limitées sur la destiné du béryllium dans l'environnement et notamment au sujet des effets qu'il exerce sur les organismes aquatiques et terrestres. Les concentrations en béryllium dans les eaux superficielles (de l'ordre du μg/litre) et dans les sols (de l'ordre du μg/kg de poids sec) sont généralement faibles et n'ont probablement pas d'effets nocifs sur l'environnement.

RESUMEN Y CONCLUSIONES

1. Identidad, propiedades físicas y químicas, métodos de análisis

El berilio es un metal quebradizo de color gris acero, cuyo único isótopo natural es el ^{9}Be. Sus compuestos son bivalentes. El berilio posee varias propiedades excepcionales. Es la más liviana de todas las sustancias sólidas y químicamente estables y tiene una temperatura de fusión, un calor específico, un calor de fusión y una resistencia por relación al peso excepcionalmente elevados. Posee excelentes propiedades de conductividad y conductibilidad. Debido a su pequeño número atómico, el berilio es muy permeable a los rayos X. Sus propiedades nucleares incluyen la ruptura, dispersión y reflexión de neutrones, así como la emisión de neutrones por bombardeo-α.

El berilio tiene una serie de propiedades químicas en común con el aluminio, especialmente su gran afinidad por el oxígeno. Esta hace que en la superficie del berilio metálico y de las aleaciones de berilio se forme una película muy estable de óxido de berilio (BeO) que los hace muy resistentes a la corrosión, el agua y los ácidos oxidantes en frío. Cuando se inflama en oxígeno, el polvo de berilio arde a una temperatura de 4500 °C. El óxido de berilio sinterizado ("berilia") es muy estable y posee propiedades cerámicas. Las sales catiónicas de berilio se hidrolizan en agua y reaccionan formando hidróxidos insolubles o complejos hidratados, cuando los valores de pH varían entre 5 y 8, y berilatos cuando el pH es superior a 8.

El berilio, como aditivo para aleaciones, confiere una combinación de propiedades notables a otros metales, especialmente la resistencia a la corrosión, un gran módulo de elasticidad, características no magnéticas y no pirofóricas, mayor conductividad y conductibilidad, además de una resistencia superior a la del acero.

Se han utilizado diversos métodos analíticos para determinar la presencia de berilio en diferentes medios. Los métodos más antiguos incluyen técnicas de espectroscopia, fluorimetría y espectrofotometría. La espectrometría de absorción atómica sin llama y la cromatografía de gases son los métodos de elección; los límites de detección son de 0,5 ng/muestra (absorción atómica sin llama) y de 0,04 pg/muestra (cromatografía de gases con detección por captura de electrones). Además, cada vez se emplea más la espectrometría de emisión atómica de plasma acoplado por inducción.

2. Fuentes de exposicion humana y ambiental

El berilio es el 35° elemento más abundante en la corteza terrestre, con un contenido medio de unos 6 mg/kg. Aparte de las gemas, la esmeralda (berilo que contiene cromo) y la aguamarina (berilo que contiene hierro), sólo 2 minerales de berilio tienen importancia económica. El berilo contiene hasta un 4% de berilio y se extrae en la Argentina, el Brasil, la India, China, Portugal, la URSS y en varios países de Africa meridional y central. La bertrandita se ha convertido en la principal fuente de este metal en los EE.UU., pese a que contiene menos del 1% de berilio.

La producción mundial anual de minerales de berilio en 1980–1984 fue de aproximadamente 10 000 toneladas, lo que corresponde a unas 400 toneladas de berilio. Pese a las considerables fluctuaciones en la oferta y la demanda de berilio debidas a esporádicos programas gubernamentales en armamento, energía nuclear e industrias aeroespaciales, se preveía, en 1986, que la demanda dé berilio aumentaría en un promedio anual de un 4% hasta 1990.

Por lo general, las emisiones de berilio durante su producción y utilización son poco importantes comparadas a las emisiones que ocurren durante la combustión de carbón y el fuel, que poseen un contenido natural medio de 1,8–2,2 mg Be/kg de peso seco y hasta 100 µg Be/litro, respectivamente. La emisión de berilio por utilización de combustibles fósiles representó aproximadamente el 93% de la emisión total de berilio en los EE.UU., uno de los principales países productores. Si se mejoran las medidas de control de las emisiones, podrá reducirse considerablemente la emisión de berilio de las centrales termoeléctricas.

Si bien la utilización de combustibles fósiles determina la concentración general de berilio en la atmósfera, las fuentes relacionadas con la producción pueden dar lugar a concentraciones ambientales localmente elevadas, especialmente donde las medidas de control son insuficientes. Igualmente, las emisiones producidas por la experimentación y el uso de cohetes impulsados por berilio podrían tener gran importancia a nivel local. La exposición ocupacional ocurre sobre todo durante el procesado de minerales de berilio, berilio metálico, aleaciones con berilio y óxido de berilio. Los únicos países con industrias productivas son el Japón, los EE.UU. y la URSS. En otros países, el metal puro, las aleaciones o el óxido de berilio cerámico importados son transformados en productos finales.

La mayor parte de los desechos del berilio provienen de medidas anticontaminantes y son reciclados o enterrados. El reciclado de la mayoría de los productos finales no es rentable debido a su pequeño volumen y bajo contenido en berilio.

Aproximadamente el 72% de la producción mundial de berilio se utiliza en forma berilio-cobre y otras aleaciones en las industrias aeroespacial, electrónica y mecánica. Alrededor de un 20% se utiliza como metal libre sobre todo en las industrias aeroespacial, de armamento y nuclear. El resto se emplea como óxido de berilio para aplicaciones cerámicas, principalmente en electrónica y microelectrónica.

3. Transporte, distribución y transformación en el medio ambiente

Los datos sobre la suerte que corre el berilio en el medio ambiente son limitados. Las partículas de óxido de berilio atmosférico regresan a la tierra por sedimentación húmeda y seca. Dentro de los valores del pH ambiental, entre 4 y 8, los minerales sedimentarios finamente dispersos fijan el berilio, evitando así que pase a las aguas subterráneas.

Parece que el berilio no se biomultiplica en absoluto en las cadenas alimentarias. La mayoría de las plantas absorben berilio del suelo en pequeñas cantidades, y sólo una parte ínfima pasa de las raíces a otras partes de la planta.

4. Concentraciones ambientales y exposición humana

Las concentraciones de berilio en las aguas superficiales y de bebida suelen ser de unos pocos µg/litro. Las concentraciones en los suelos varían entre 1 y 7 mg/kg. Por lo general, las plantas terrestres contienen menos de 1 mg de berilio por kg de peso seco. En diversos organismos marinos se han encontrado concentraciones de aproximadamente 100 µg/kg de peso en fresco.

Se observaron variaciones de la concentración de berilio atmosférico en zonas rurales de los EE.UU. de 0,03 a 0,06 ng/m^3. En los países donde se queman menos combustibles fósiles es probable que las concentraciones ambientales sean inferiores. Se observó que las concentraciones medias anuales de berilio en el aire urbano de los EE.UU. variaban entre < 0,1 y 6,7 ng/m^3. En las ciudades japonesas se encontró un promedio de 0,04 ng/m^3, con valores máximos (0,2 ng/m^3) en las zonas industriales.

Antes de que se establecieran las medidas de control en los años cincuenta, las concentraciones de berilio en la atmósfera eran sumamente elevadas en las cercanías de las plantas de producción y procesamiento. Además, se daban frecuentes casos de exposición "paraocupacional" en las familias de los trabajadores, denominados casos de vecindad, debidos al contacto con la ropa del trabajador, o a la exposición atmosférica, o a ambos. Hoy en día, esas fuentes de exposición suelen ser insignificantes para la población en general. La fuente principal de exposición ambiental de la población general al berilio atmosférico es el empleo de combustibles fósiles. Excepcionalmente, también puede haber una exposición elevada en las cercanías de centrales eléctricas donde se queme carbón con altas concentraciones de berilio y no se apliquen las medidas de control adecuadas. Probablemente, fumar tabaco también sea una fuente importante de exposición al berilio.

El creciente uso del berilio en la base de aleaciones para piezas dentales podría tener cierta importancia para la población general, debido al gran potencial del berilio de provocar reacciones alérgicas por contacto.

Antes de 1950, la exposición al berilio en los lugares de trabajo era a menudo muy elevada; no era raro encontrar concentraciones superiores a 1 mg/m^3. Las medidas de control establecidas en distintos países para responder a las normas ocupacionales de 1-5 μg Be/m^3 (promedio ponderado por el tiempo) redujeron drásticamente dichas concentraciones, aunque todavía no se han alcanzado estos valores en todas partes.

La presencia de berilio en tejidos y líquidos orgánicos puede ser indicativa de una exposición anterior. Las personas sin exposición específica presentan valores en la orina de alrededor de 1g/litro y en el tejido pulmonar inferiores a 20 g/kg (peso seco). Los datos limitados de que se dispone no permiten establecer una clara relación entre exposición y concentración en el cuerpo, si bien se han encontrado valores elevados (20 g/kg) en muestras de tejido pulmonar de pacientes con la enfermedad del berilio.

5. Cinética y metabolismo

No se dispone de datos sobre el depósito o la absorción del berilio inhalado en el hombre. Los estudios en animales han demostrado que, tras depositarse en los pulmones, el berilio permanece en ellos y es absorbido lentamente en la sangre. La capacidad de autodepuración pulmonar es bifásica, con una fase de eliminación rápida durante las primeras 1-2 semanas después de haber cesado la exposición.

La mayor parte del berilio que circula en la sangre es transportado en forma de fosfato coloidal. Una parte importante de la dosis inhalada se incorpora en el esqueleto, siendo éste el lugar final donde se almacena el berilio. Generalmente, la exposición por inhalación tiene como consecuencia un almacenamiento a largo plazo de cantidades apreciables de berilio en el tejido pulmonar, particularmente en los nódulos linfáticos pulmonares. Los compuestos más solubles de berilio son también transportados al hígado, los nódulos linfáticos abdominales, el bazo, el corazón, el músculo, la piel y el riñón.

Por lo común, tras la administración oral de berilio, una pequeña cantidad (menos del 1%) era absorbida en la sangre y almacenada en el esqueleto. También se encontraron pequeñas cantidades en el tracto gastrointestinal y en el hígado.

La absorción de berilio por la piel intacta es insignificante, puesto que los constituyentes de la epidermis fijan el berilio.

Una proporción considerable del berilio absorbido se elimina rápidamente, sobre todo por la orina y, en cierta medida, por las heces. Una parte del berilio inhalado se elimina en las heces, probablemente como resultado de la capacidad de autodepuración del tracto respiratorio y la ingestión de berilio por vía oral.

Debido al prolongado almacenamiento del berilio en el esqueleto y los pulmones, su semivida biológica es sumamente larga. En el caso del hombre se ha calculado que la semivida permanece 450 días en el esqueleto.

6. Efectos en los organismos del medio ambiente

Los microorganismos del suelo cultivados en un medio deficiente en magnesio crecen mejor en presencia de berilio, debido a la sustitución parcial del magnesio por el berilio en el metabolismo de los organismos. Se observaron efectos similares de estimulación del crecimiento en algas y en plantas cultivadas. Al parecer este fenómeno depende del pH, ya que ocurre únicamente cuando el pH es elevado. Cuando el pH es igual o inferior a 7, el berilio es tóxico para las plantas acuáticas y terrestres, sean cuales sean las concentraciones de magnesio en el medio de crecimiento.

En general, los compuestos de berilio solubles inhiben el crecimiento de las plantas en concentraciones de mg/litro. Por ejemplo, en habichuelas (*Phaseolus vulgaris*) cultivadas en una solución nutritiva con un pH 5,3, se observó una reducción del rendimiento del 88% con una concentración de 5 mg Be/litro. Los efectos se observaron en primer lugar en las raíces, que se oscurecieron y dejaron de elongarse normalmente. Las raíces acumulan la mayor parte del berilio absorbido y sólo una pequeña cantidad es transportada a las partes superiores de la planta. Se estimó que el contenido crítico de berilio que reduce un 50% el rendimiento es de unos 3000 mg Be/kg en las raíces y de unos 6 mg Be/kg en las hojas externas de las plantas de col (*Brassica oleracea*) en base al peso seco.

En alubias, trigo y trébol ladino cultivados en tierra se observó atrofia radicular y foliar, pero no clorosis ni manchas en las hojas.

En los cultivos en tierra, la fitotoxicidad del berilio está regida por las características del suelo, en particular su capacidad de intercambio catiónico y el pH de la solución del suelo. Aparte del efecto de sustitución del magnesio, la disminución de la fitotoxicidad en condiciones alcalinas también se debe a la precipitación del berilio en forma no disponible como sal de fosfato.

El mecanismo responsable de la fitotoxicidad del berilio se basa probablemente en la inhibición de enzimas específicas, especialmente las fosfatasas vegetales. El berilio también inhibe la absorción de iones minerales esenciales.

En estudios de toxicidad aguda con diferentes especies de peces de agua dulce, se observaron valores de CL_{50} de 0,15 a 2 mg Be/litro, según la especie y las condiciones experimentales. La toxicidad para los peces aumentaba cuando disminuía la dureza del agua; el sulfato de berilio era más tóxico en uno o dos órdenes de magnitud para *Pinephales promelas* y *Leponis macrochirus* en agua blanda que en agua dura. Las larvas de salamandra y la pulga de agua *Daphnia magna* mostraron una sensibilidad similar.

No existen datos validados sobre la toxicidad crónica del berilio en animales acuáticos, si bien en un estudio inédito se observó que *Daphnia magna* se veía afectada adversamente por concentraciones de berilio bastante menores (5 μg Be/litro) en las pruebas de reproducción a largo plazo que en las pruebas de toxicidad aguda (CE_{50} 2500 μg Be/litro).

7. **Efectos en animales de experimentación y en sistemas de ensayo *in vitro***

    Los síntomas de envenenamiento agudo con berilio que se manifestaron en los animales de experimentación fueron trastornos respiratorios, espasmos, choque hipoglucémico y parálisis respiratoria.

    La implantación de compuestos de berilio y de berilio metálico en los tejidos subcutáneos puede producir granulomas similares a los observados en el ser humano. En el cobayo apareció hipersensibilidad tras la inyección de compuestos solubles de berilio por vía intradérmica.

    Como efecto secundario, el carbonato de berilio produjo raquitismo en ratas jóvenes por la precipitación intestinal de fosfato de berilio y la correspondiente privación de fósforo.

Varias especies animales presentaron neumonitis químicas agudas tras la inhalación de metal de berilio o de diferentes compuestos de berilio, incluso las formas insolubles. Las exposiciones diarias repetidas a vahos de sulfato de berilio, de una concentración media de 2 mg Be/m^3, fueron letales para la rata (90% de muertes), el perro (80%), el conejo (10%), el cobayo (60%), el mono (100%), la cabra (100%), el hámster (50%) y el ratón (10%). Debido al efecto sinérgico del ion fluoruro, los efectos del fluoruro de berilio fueron unas dos veces superiores a los de los sulfatos. Algunas de las lesiones en los pulmones eran parecidas a las observadas en el hombre, pero los granulomas no eran idénticos.

La toxicidad de la inhalación de óxido de berilio insoluble depende en gran parte de sus propiedades físicas y químicas, que pueden variar considerablemente según las condiciones de producción. Debido a que el tamaño mínimo de las partículas es más pequeño y la agregación es menor, la exposición a BeO caldeado a baja temperatura (400 °C) a 3,6 mg Be/m^3 durante 40 días fue causa de mortalidad en ratas y de graves lesiones pulmonares en perros. El BeO caldeado a dos temperaturas elevadas distintas (1350 °C y 1150 °C) no causó lesiones pulmonares, pese a que la exposición total fue mayor (32 mg Be/m^3, 360 h).

La respuesta no maligna característica a la exposición a largo plazo por inhalación de concentraciones menores de compuestos de berilio solubles e insolubles es una neumonitis crónica asociada a granulomas, que sólo corresponde en parte a la enfermedad crónica en el ser humano.

Los resultados de las pruebas de genotoxicidad indican que el berilio interacciona con el ADN y causa mutaciones de genes, aberraciones cromosómicas y un intercambio de cromátidas hermanas en cultivos de células somáticas de mamíferos, aunque no presenta efectos mutagénicos en sistemas bacterianos de ensayo.

La inyección intravenosa (3,7–700 mg Be) e intramedular (0,144–216 mg Be) de berilio metálico y de diversos compuestos produjo osteosarcomas y condrosarcomas en conejos, con la aparición de metástasis en un 40–100% de los animales, especialmente en el pulmón.

En ratas, la inhalación (0,8–9000 μg Be/m^3) o la exposición intratraqueal (0,3–9 mg Be) a compuestos solubles o insolubles de berilio, berilio metálico,

y diversas aleaciones de berilio indujo tumores pulmonares del tipo del adenoma o del adenocarcinoma, parcialmente metastatizante. El berilio (620 μg Be/m^3) fue el único mineral de berilio que produjo carcinomas pulmonares (no así la bertrandita a 210 μg Be/m^3). El óxido de berilio resultó ser carcinogénico para la rata, pero la incidencia de adenocarcinomas pulmonares fue mucho mayor tras la administración intratraqueal (9 mg Be) de una especificación caldeada a baja temperatura (51%), en comparación con la de óxidos caldeados a alta temperatura (11–16%). En la época en que se realizaron muchos de estos estudios, la concepción de los estudios y las prácticas de laboratorio no solían ajustarse a las prácticas actuales, por lo que conviene considerar con especial precaución los datos comunicados sobre inhalación.

La inducción de cáncer del pulmón por el berilio varía mucho de unas especies a otras. Mientras que la rata y quizá el mono son muy susceptibles a este respecto, no se han observado tumores pulmonares en el conejo, el hámster ni el cobayo.

Existen tres teorías en cuanto a los mecanismos de toxicidad del berilio: 1)el berilio incide en el metabolismo del fosfato ya que inhibe enzimas cruciales, en particular la fosfatasa alcalina; 2)el berilio inhibe la replicación y la proliferación celular ya que afecta a las enzimas del metabolismo de los ácidos nucleicos; y 3)en la toxicidad del berilio interviene un mecanismo inmunológico, como se ha observado en el cobayo, que desarrolla una hipersensibilidad cutánea de mediación celular.

8. Efectos en el hombre

La exposición al berilio de importancia toxicológica se limita casi exclusivamente al lugar de trabajo. Antes de la introducción de medidas mejores de control de la emisión y de higiene en las plantas de berilio, se registraron varios casos de "vecindad" de la enfermedad crónica del berilio. Hasta 1966 se habían registrado un total de 60 casos en los EE.UU., algunos de ellos por contacto con la ropa de los trabajadores (exposición "paraocupacional") o exposición a la atmósfera en las cercanías de las plantas de berilio. En los últimos años no se ha notificado ningún caso.

Recientemente, se han notificado varios casos de estomatitis alérgica por contacto, ocasionada probablemente por prótesis dentales que contienen berilio.

En los años treinta y cuarenta hubo varios centenares de casos de enfermedad aguda del berilio, sobre todo entre los trabajadores de las plantas de extracción de berilio en Alemania, Italia, los EE.UU. y la URSS. La inhalación de sales solubles de berilio, especialmente el fluoruro y el sulfato, en concentraciones superiores a 100 μg Be/m^3, produjo síntomas agudos en casi todos los trabajadores expuestos, mientras que con concentraciones iguales o menores a 15 μg/m^3 (determinadas por métodos analíticos anticuados), no se registró ningún caso. Tras la adopción a principios de los años cincuenta de la concentración máxima de exposición de 25 μg/m^3, hubo una disminución drástica de los casos de la enfermedad aguda del berilio.

Los signos y síntomas de la enfermedad aguda del berilio varían desde una leve inflamación de las mucosas nasales y la faringe hasta una traqueobronquitis y neumonitis química grave. En los casos graves, los pacientes fallecieron de neumonitis aguda, pero en la mayoría de los casos, al cesar la exposición se produjo una recuperación total en el lapso de 1–4 semanas. En algunos casos, la enfermedad crónica del berilio se desarrolló años después de la recuperación de la forma aguda de la enfermedad.

El contacto directo con compuestos solubles de berilio produce dermatitis por contacto y posiblemente conjuntivitis. Los individuos sensibilizados reaccionan mucho antes y a cantidades menores de berilio. La introducción cutánea o subcutánea de compuestos de berilio solubles o insolubles produce ulceraciones crónicas; a menudo aparecen granulomas al cabo de varios años.

La enfermedad crónica del berilio difiere de la forma aguda en que tiene un periodo de latencia que varía desde unas semanas hasta más de 20 años; es además de larga duración y de gravedad progresiva. En el registro oficial central de casos de enfermedad del berilio de los EE.UU., creado en 1952, se habían registrado 888 casos hasta 1983. Seiscientos veintidós se clasificaron como

Resumen

crónicos, de los cuales 557 se debían a la exposición ocupacional, sobre todo en la industria de lámparas fluorescentes (319 casos) o en las plantas de extracción de berilio (101 casos). A partir de 1949, cuando se abandonó el uso del silicato de berilio y zinc y del óxido de berilio en los fósforos de tubos fluorescentes y se adoptó un límite de exposición ocupacional (TWA, 2 μg Be/m^3), el número de casos de enfermedad crónica del berilio disminuyó ostensiblemente, si bien se han registrado nuevos casos como consecuencia de la exposición a una concentración en el aire de unos 2 μg/m^3.

El término "enfermedad crónica del berilio" se prefiere al de "beriliosis" debido a que esta enfermedad difiere de la típica neumoconiosis. Las características más típicas son la inflamación granulomatosa del pulmón, asociada a una disnea tras un esfuerzo, tos, dolor de pecho, pérdida de peso, fatiga y debilidad general; también puede darse una hipertrofia del corazón derecho con el consiguiente fallo cardiaco, hepatomegalia, esplenomegalia, cianosis y dedos en palillo de tambor. Asimismo, se han observado, asociados a la enfermedad crónica del berilio, cambios en las proteínas séricas y la función hepática, cálculos renales y osteoesclerosis. La evolución de la enfermedad crónica del berilio no es uniforme; en algunos casos se observa una remisión espontánea durante semanas o años, seguida de exacerbaciones. En la mayoría de los casos, se observa una enfermedad pulmonar progresiva con un mayor riesgo de muerte por fallo cardiaco o respiratorio. Se han notificado tasas de morbilidad entre los trabajadores del berilio que varían entre 0,3 y 7,5%. En los pacientes con enfermedad crónica del berilio las tasas de mortalidad alcanzan hasta el 37%.

A nivel macroscópico los pulmones pueden presentar cambios difusos, con pequeños nódulos muy dispersos y fibrosis intersticial. A nivel microscópico, existen granulomas de tipo sarcoide con diferentes grados de inflamación intersticial, que generalmente no pueden diferenciarse de los observados en otras granulomatosis como la sarcoidosis o la tuberculosis.

Para diagnosticar la enfermedad crónica del berilio, son de gran utilidad el historial y un análisis de tejido, aunque la presencia de berilio en el material biológico no constituye una prueba de que haya enfermedad. Las pruebas alérgicas no son recomendables, dado que no son muy fiables y ellas mismas producen una alta sensibilización. Los elementos más útiles para el diagnóstico son el ensayo de inhibición de la migración de macrófagos y la prueba de transformación de linfocito-blastos.

Estos métodos que miden la hipersensibilidad se basan en un mecanismo inmune que probablemente es subyacente a la enfermedad crónica del berilio y la tardía hispersensibilidad cutánea y granulomatosa.

En la enfermedad crónica del berilio, la sensibilización inmunológica podría explicar la gran variabilidad de la latencia y la falta de relación entre la dosis y la respuesta. El embarazo parece ser un "factor de estrés" desencadenante, puesto que el 66% de las 95 mujeres registradas entre los casos mortales del registro de casos de enfermedad del berilio en los EE.UU. estaban embarazadas.

Las fuentes de exposición de los pacientes con la enfermedad del berilio incluyen la producción de aleaciones metálicas de berilio, maquinarias, la investigación y producción de cerámicas, y la producción de energía. Es posible que las actuales normas de exposición ocupacional no basten para impedir la aparición de la enfermedad crónica del berilio en individuos sensibilizados.

La carcinogenicidad del berilio ha sido examinada en varios estudios realizados sobre los trabajadores empleados en dos instalaciones de producción de berilio en los EE.UU. y en un registro de casos clínicos de afecciones pulmonares relacionadas con el berilio; el registro procedía de estas instalaciones y de otras ocupaciones. Los resultados de estos estudios fueron puestos en tela de juicio debido al sesgo en la selección, a la confusión con los efectos de fumar cigarrillos y a que se subestimó el número de muertes previstas por cáncer del pulmón, dado que se utilizaron las tasas de mortalidad correspondientes al periodo 1965–1967 para hacer una estimación de la mortalidad prevista para los años 1968–1975. Si bien es poco probable que los dos primeros factores desempeñen un papel importante en el exceso de riesgo de cáncer del pulmón, los datos que se dan en este documento se basan en un "ajuste" del número de muertes previstas por cáncer del pulmón. Todos los estudios realizados señalaban riesgos significativamente elevados de cáncer del pulmón.

9. Evaluación de los riesgos para la salud humana y los efectos en el medio ambiente

9.1 Riesgos para la salud humana

Siempre y cuando las medidas de control en la industria del berilio sean adecuadas, la exposición de la población general hoy en día se limita a pequeñas concentraciones de berilio en la atmósfera, procedente de la utilización de combustibles fósiles. En casos excepcionales, cuando se quema carbón con un contenido de berilio desusadamente alto, pueden surgir problemas de salud. Habría que examinar nuevamente el uso de berilio en las prótesis dentales debido a su alto potencial de sensibilización.

Los casos de enfermedad aguda del berilio que producen nasofaringitis, bronquitis y neumonitis química grave han disminuido notablemente, y hoy en día sólo podrían producirse a consecuencia de errores en los sistemas de medidas de control. La enfermedad crónica del berilio se diferencia de la forma aguda en que tiene un periodo de latencia que varía entre unas semanas y más de 20 años, es de larga duración y de gravedad progresiva. Principalmente afecta al pulmón. La característica típica es la inflamación granulomatosa del pulmón asociada a disnea por esfuerzo excesivo, tos, dolor de pecho, pérdida de peso y debilidad general. Los efectos en los otros órganos pueden ser secundarios en lugar de sistémicos. La gran variabilidad de la latencia y la falta de relación dosis-respuesta pueden observarse aún hoy en día en individuos sensibilizados que han estado expuestos a una concentración de unos 2 $\mu g/m^3$.

Pese a algunas deficiencias en la concepción de los estudios y las prácticas de laboratorio, la actividad carcinogénica del berilio en diferentes especies animales ha sido confirmada.

Varios estudios epidemiológicos han demostrado que existe un riesgo excesivo de cáncer del pulmón debido a la exposición ocupacional al berilio. Pese a las críticas sobre la interpretación de estos resultados, con los datos disponibles se llega a la conclusión de que el berilio es la explicación única más probable del exceso de cáncer del pulmón en los trabajadores expuestos.

9.2 Efectos en el medio ambiente

Los datos sobre la suerte que corre el berilio en el medio ambiente, incluso sus efectos en organismos acuáticos y terrestres, son limitados. La concentración de berilio en las aguas superficiales (orden de g/litro) y los suelos (orden de mg/kg de peso seco) es generalmente pequeña y es probable que no afecte negativamente al medio ambiente.

www.ingramcontent.com/pod-product-compliance
Ingram Content Group UK Ltd.
Pitfield, Milton Keynes, MK11 3LW, UK
UKHW021312180426
11947UKWH00015B/1186